THE STORY OF THE OXFORDSHIRE & BUCKINGHAMSHIRE LIGHT INFANTRY

The Story of the

OXFORDSHIRE AND BUCKINGHAMSHIRE LIGHT INFANTRY

(The old 43rd and 52nd Regiments)

BY

SIR HENRY NEWBOLT

CONTENTS

CHAPTER I

WITH WOLFE AT QUEBEC, 1741–1759

CHAPTER II

FALL OF FRENCH EMPIRE IN CANADA, 1759–1764

CHAPTER III

AMERICAN WAR OF INDEPENDENCE, 1765–1782

CHAPTER IV

FIGHTING IN EAST AND WEST INDIES, 1782–1799

CONTENTS (*continued*)

CONTENTS (*continued*)

CHAPTER IX

THE PENINSULA: VITTORIA TO THE NIVE, 1813

CHAPTER X

CLOSE OF PENINSULAR CAMPAIGN, 1814

CHAPTER XI

THE AMERICAN WAR, 1815

CHAPTER XII

WATERLOO, 1815

CONTENTS (*continued*)

CHAPTER XIII

THE INDIAN MUTINY, 1809–1859

CHAPTER XIV

NEW ZEALAND, TIRAH, AND SOUTH AFRICA, 1859–1914

CHAPTER XV

THE GREAT WAR, 1914

APPENDICES

LIST OF ILLUSTRATIONS

PLATES

FIGURES IN TEXT

LIST OF ILLUSTRATIONS (*continued*)

A Private of the 52nd.

A Private of the 43rd. 1742

STORY OF
THE OXFORDSHIRE AND BUCKINGHAMSHIRE LIGHT INFANTRY

CHAPTER I

1741–1759

The record of a great regiment—The quality of soldierliness—The predestined union of the 43rd and 52nd—Wolfe and the 43rd at Quebec.

SINCE a regiment is at once an association of men and a unit of the national force, its record will be partly biographical and partly historical: it will touch upon many moments of intensest feeling, and will always be following the course of great events. With such material it should make a stirring book. However personal the story may be at times, it can hardly be trivial: the adventures, escapes and promotions of individuals—even their mistakes and miseries, their oaths and jests—these in war are something more than anecdotes, for the smallest act becomes significant when done in presence of death. On the other hand, however historical the narrative may be, it can hardly be unsympathetic, for the army's motive is that of all good patriots and is never in doubt for a moment, so that its every effort calls to our sense of fellowship.

But if our sympathy is to be perfect we must be given at least a hint of a third element in the story: we must remember not only the soldier's object, but the means by which he attempts to gain it. A battle is far more than a historical event or a series of personal adventures: it is an example of the art of war. This may sound at first a duller business, but it is not so. War depends for its principles on science, but for its practice on high moral virtues. The most important of these is not, as many think, either courage or comradeship. "Splendid fellows," say our generals of their young officers, "but the rarest thing even now is to find a soldier among them." That rare virtue of the soldier is the resignation of all ends but one, and the continual concentration of the will and the intelligence upon that one. Such a power may be with some men a natural gift: for the majority it is attainable only by long and devoted training, and nowhere but in the school of a great regiment, inspired and sustained by a great tradition.

The result of such a training is visible at any distance of time. To us, as to their contemporaries, the regiments of the Light Division prove continually the genius of the men who made them—the genius of Wolfe, the genius of Sir John Moore. We see them move across the fields of the Peninsula with a kind of heroic majesty: they know their work, they see what they foresaw, their fighting has the grandeur of conscious mastery. The game itself, not the glory of it, is the secret they have handed on to their successors. For the men of this tradition fame is inevitable, but it is no more than the light upon their steel. Their reward is the trust of their commanders and the victory of the cause they serve.

The regiment now known as the Oxfordshire and Buckinghamshire Light Infantry has for its 1st and 2nd battalions the old 43rd and 52nd of the line. The union of these two seems to have been ordained by fate from the beginning. The first omen was a coincidence of number. The older battalion of the two, raised in
1741 as the 54th, was renumbered seven years later as 1741
the 43rd: and by mere chance the younger battalion, raised in 1755, also bore for the first two years of its existence that same title of the 54th, until, on the disbanding of Shirley's and Pepperell's regiments, it became the 52nd. The fulfilment of the omen began in 1774, when the 43rd and 52nd took part in the American War. They charged side by side at Bunker's Hill, they served together in Howe's campaigns: in 1807 they went together on the Copenhagen expedition: from 1808 they marched and fought together through the seven years of the Peninsular War. Their remarkable unity of historic service was finally commemorated in 1881, when they were linked together in one regiment by the name of the Oxfordshire Light Infantry.

But close as this comradeship has been, it has, of course, not been continuous. During the 150 years of their existence there have been many wars in which only one of the two regiments was engaged. Of the two most momentous battles in our history, each battalion was present at one and absent from the other: the 52nd were at Waterloo, the 43rd at Quebec. It is with Quebec that our story must begin.

On May 8, 1757, a force of over 5000 men sailed 1757
from Cork for the defence of the British North Ameri-

1757 can Colonies. The 43rd, 700 strong, filled six of the forty-five transports, and the expedition, convoyed by a fleet of fourteen ships of the line under Admiral Holborne, arrived off Halifax at the end of June. For nearly two years the 43rd spent their time uneventfully, garrisoning Fort Cumberland, Annapolis Royal and Fort Edward, living on salt rations and sea biscuits, apples and spruce beer, skirmishing with the French and Indians, and training eagerly for trench warfare in hopes of more active service.

1759 In April 1759 a triple attack upon the French in Canada was ordered. General Amherst, the Commander-in-Chief, was to march on Ticonderoga, General Prideaux on Niagara and Montreal, and General Wolfe, supported by the fleet under Admiral Saunders, was to take 8000 men up the St. Lawrence and besiege Quebec. This force concentrated at Louisbourg, where the 43rd joined on May 24. They were at once reviewed by General Wolfe and set to work at his new system of drill. Their heavy firelocks were exchanged for fusils, and spades, pickaxes, shovels, and bill-hooks were served out.

On the 5th and 6th of June the expedition got under weigh, convoyed by a fleet of twenty-two of the line, five frigates and sixteen sloops. Captain the Hon. John Knox, of the 43rd, who has left an account of the campaign, observes with "inexpressible pleasure" the high spirits of the whole armament, and records "the prevailing sentimental toast among the officers to be 'British colours on every French fort, port and garrison in America!'" In the fortunes of war sweeping ambitions are seldom justified: but this one was destined to complete and lasting fulfilment.

On June 16 the 43rd found themselves in the mouth

of the St. Lawrence : on the 19th General Wolfe passed 1759
them in the *Richmond* frigate : on the 27th they landed on the Isle of Orleans, which lies in midstream and divides the St. Lawrence just where the channel widens suddenly below Quebec. From the Point of Orleans, at the western or upstream end of the island, the whole French position was visible. It stretched along the north shore of the river : first the city of Quebec on its high, jutting promontory, then on the east of it the broad mouth of the river St. Charles, then still further to the right the shoals of Beauport, with a coastline of some eight miles defended by earthworks, the eastern flank of which was covered by the Falls of Montmorency. For the defence of this ten-mile line Vaudreuil, the Governor of Quebec, and Montcalm and Levis, the French Generals, had 15,000 troops and some Indians : for the attack Wolfe had less than 9000.

But to say this is to leave the sailors out of account : and they had already done enough to gain the confidence of the army. When the French pilot on Captain Knox's transport declared that the ship could not pass the Isle of Orleans, " Aye, aye, my dear ! " replied old Killick, the master, who was steering her, " but damme, I'll convince you that an Englishman shall go where a Frenchman dare not show his nose." And Vaudreuil himself had the mortification of informing his government that " The enemy have passed sixty ships of war where we daren't risk a vessel of a hundred tons by night or day." It was this mobility, the power to move along the whole French front and strike where he chose, that made Wolfe's victory possible.

The French were not unaware of this danger, and they made an immediate attempt to counter it. At

1759 midnight on the 28th they sent down five fireships and two rafts to destroy the British fleet; but some of these drifted ashore, and the rest were towed clear by our sailors. "They were certainly," says Captain Knox, "the grandest fireworks (if I may be allowed to call them so) that can possibly be conceived, every circumstance having contributed to their awful yet beautiful appearance: the night was serene and calm, there was no light but what the stars produced, and this was eclipsed by the blaze of the floating fires, issuing from all points, and running almost as quick as thought up the masts and rigging: add to this the solemnity of the sable night, still more obscured by the profuse clouds of smoke, with the firing of the cannon, the bursting of the grenades and the crackling of the other combustibles: all which reverberated through the air and the adjacent woods, together with the sonorous shouts, and frequent repetitions of *All's Well* from our gallant seamen on the water."

The Point of Orleans was two and a half miles from Quebec: in order to bring his guns within effective range Wolfe made a landing at Point Levi, a height on the south side of the river and immediately opposite to the lower part of the city. The position was occupied on the 29th and 30th by the 43rd and other regiments of Monckton's Brigade: earthworks were thrown up, and on July 12 the batteries opened fire. This bombardment was damaging to the town, but not to the French army; and Wolfe soon determined to attempt the other end of the enemy's long line. He threw Murray's and Townshend's Brigades across to the north shore, and entrenched them in a camp close to the gorge of Montmorency, which he hoped to be able to force. On the 18th he sent part of the fleet

upstream, past the batteries of Quebec, and so drew 1759
off some of Montcalm's troops to the west. On the 27th the sailors off the Isle of Orleans warded off a second fleet of fireships.

On July 31 Wolfe made his first attack and failed. His plan was to throw Monckton's Brigade ashore at Point de Leste, just inside the extreme French left, and Townshend, with 2000 men from the new camp, was to cross the Montmorency below the Falls and support the movement against the higher ground. The General himself embarked at Point Levi, taking with him the 15th and the 78th (leaving the 43rd ready in reserve), and the grenadiers of the whole army, including those of the 43rd. The *Centurion*, a 64-gun ship, covered the landing, which was accomplished under a hot fire: Wolfe himself was three times touched by splinters, and a cannon ball carried away the cane from his hand. The French abandoned an outlying redoubt and retired within their main entrenchments. Townshend began to move across the Montmorency, and all appeared to be going well.

At this moment the whole scheme was ruined by the impetuosity of the grenadiers. Excited by the capture of the redoubt, they rushed at the hill without waiting for orders or giving time for their supports to follow: their officers went with them because they could not hold them back. A heavy rainstorm and the hail of French bullets broke upon them at the same instant, and after a futile exhibition of undisciplined valour they were beaten back to the beach, leaving many of their wounded to the tomahawks and scalping knives of the Indians. Their total loss was four officers killed and twenty-nine wounded, with more than 400 of the rank and file, out of a force of about 1000 men.

1759 The grenadiers of the 43rd lost nine killed, and two officers and thirteen men wounded.

It was too late in the day for such a repulse to be made good: Wolfe, with perfect self-command, re-embarked his troops at once. That night he visited every wounded grenadier officer, and invited all those who were unwounded to dine with him. The men he severely rebuked next day for their "impetuous, irregular and unsoldierlike proceedings." He then began to prepare for fresh operations by sending Murray's division with the fleet to destroy the French ships above Quebec.

His next difficulty was an epidemic of fever, which sent 1000 of his men into hospital. By August 20 he himself was down, and on the 22nd the army heard "with the gravest concern" that he was very ill. On the 25th Captain Knox reports him to be "on the recovery, to the inconceivable joy of the whole army." On the same day came the good news of "the reduction of Niagara by a detachment of Mr. Amherst's army."

On September 2 Wolfe finished and sent to Pitt his famous dispatch on the first failure and proposed new attack. "I found myself so ill and am still so weak," he wrote, "that I begged the General Officers to consult together for the public utility. *They are all of opinion,* that as more ships and provisions are now got above the town, they should try, by conveying up a corps of four or five thousand men—which is nearly the whole strength of the army after the Points of Levi and Orleans are left in a proper state of defence—to draw the enemy from their present situation and bring them to an action. *I have acquiesced in their proposal,* and we are preparing to put it in execution."

On the following day Townshend's division was withdrawn from the camp at Montmorency. On the 4th expresses brought news of Amherst's victories at Ticonderoga and Crown Point: but Wolfe was very ill again and the army "very apprehensive lest he should not be able to command this grand enterprise in person." On the 5th he was reported better: on the 6th the move began. At three o'clock that afternoon the 15th, 43rd, and 78th regiments, with Brigadiers Monckton and Townshend, marched upstream to Goreham's Post, forded the river Etchemin under fire from a French battery at Sillery, and were embarked—some in transports, the 43rd on board the *Seahorse* frigate, where they were more crowded, but were entertained "in a most princely manner" by Captain Smith and his officers. The same night Wolfe himself went on board the 50-gun ship *Sutherland*. 1759

On the 9th the crowding on the ships was found to be unbearable: the 43rd were moved to the *Employment* transport, and 1500 of the other troops were put ashore again. On the 10th Wolfe made a reconnaissance with Townshend and his chief engineer, attended by an officer and thirty men of the 43rd as escort. During the day he chanced to overhear a conversation about two other officers of the regiment who were ill aboard their transport. "He has but a puny, delicate constitution" was the account given of one of the two. "This," says Knox, "struck his Excellency as being his own case," and he interrupted with the remark, "Don't tell me of constitution; that officer has good spirits, and good spirits will carry a man through everything." He afterwards offered to send the sick men back to Goreham's Post in his barge: but they proved the truth of his own saying by assuring him "that no

1759 consideration could induce them to leave the army until they should see the event of this expedition."

The enemy were now becoming alarmed at Wolfe's movements up and down the front of their position, and Montcalm hurried a large body of troops upstream. But Wolfe's plan was to land "where the French seem least to expect it," namely, under the high and steep cliffs close to the weak front of the city. The venture seemed desperate, but, as the General wrote in his final order, "A vigorous blow struck by the army at this juncture may determine the fate of Canada." He added, with a phrase that forestalled Nelson's signal, "The officers and men will remember what their country expects from them."

On the 12th Monckton and Murray re-embarked their troops: at nine o'clock that night the first division, numbering 1600 men, were put into flat-bottomed boats and moved upstream with the flood tide, attended by the ships of the squadron. The 43rd, in four boats, were in charge of the 60-gun ship *Pembroke*. About an hour before dawn the whole flotilla moved slowly down with the ebb, and Wolfe gave the signal for landing by showing a second light in the maintopmast shrouds of the *Sutherland*. He led the way himself, and as the boat drifted in to shore, he quietly recited Gray's "Elegy" to the officers with him. Perhaps only a soldier could recognize, upon the very threshold of action, the supremacy of the power that transfigures life and all its acts. "Gentlemen," he said, "I would sooner have written that poem than take Quebec."

The place of landing was at the Anse du Foulon, 120 yards east of Sillery and about two miles west of Quebec. French sentries fired with some effect—in one boat the 43rd had three men killed and four wounded—

but the alarm was not given to the main guard above. 1759
As fast as the troops landed the boats were sent back to bring on the supports, and the advanced guard of light infantry led the ascent at once "up one of the steepest precipices that can be conceived, being almost perpendicular and of an incredible height." By the time the 43rd reached the top it was clear daylight: all was quiet and not a shot was heard—the French sentries had been answered in their own language by Captain Donald M'Donald of the 78th Highlanders. A line was formed, and the troops moved forward with Quebec on their right and Sillery on their left, the light troops capturing the redoubts of Sillery.

Wolfe now faced his men to the right and marched towards the city until he reached the Plains of Abraham, a level plateau which he had marked as a good battle-ground while his troops were forming up after their climb. The enemy made their appearance about six o'clock, and he then formed his line of battle, with the 28th and the grenadiers of Louisbourg on the right, and the 43rd and 47th on the left, light infantry and grenadiers covering both flanks. By this time the 15th and 35th had come up and formed a second line: they were soon followed by the 48th, 58th, 78th, and two battalions of the 60th, and it became possible to make a much stronger front. The 35th were moved up on to the right of the first line, and the 78th and 28th on to the left, so that the 43rd now became the centre of the seven regiments; the 15th and 60th were in support, under Townshend, with the 48th in reserve under Colonel Burton. Wolfe, Monckton and Murray were in the front line.

About seven o'clock the enemy opened fire with a few guns: some Indians and sharpshooters also ad-

1759 vanced, but were driven off by the 47th. Our men were ordered to lie down and wait for the arrival of their own guns. These came up by eight o'clock—two short brass six-pounders—and opened fire effectively. Montcalm replied by forming his army in three large columns and attacking the left flank with his light cavalry, while Bougainville, who was coming back from the west, attempted to cut Townshend's line of communication with the landing-place. At ten o'clock, finding that both these movements had failed, Montcalm advanced his whole force briskly, two columns against Wolfe's left and one against his right. They came on with loud shouts, firing obliquely, as they advanced, into the two extremities of the British line, at a range of 130 yards. Our guns made havoc among them with grape, but the infantry were silent: Wolfe had ordered them to load with two bullets each, but to hold their fire till the word was given. The result was decisive: on this day there were no "irregular and unsoldierlike proceedings." When the French were within forty yards "a well-timed, regular and heavy discharge" broke upon them: the 43rd and 47th, in particular, who were not directly attacked, firing as if "at a private field of exercise." Every ball took effect, and a second volley was ready by the time the smoke had thinned: the French officers afterwards declared that they had never seen "such regularity and discipline—our troops, and particularly the central corps, having levelled and fired *comme un coup de canon.*"

Wolfe's hour was come: he drew his sword and led the whole line forward. The French columns gave way and were driven back on the town, the British firing as they advanced and making many prisoners. But the battle was not yet a rout: the sharpshooters were still

Phinn's Plan of Quebec.

1759 active on both flanks. Wolfe was struck first in the wrist, and then in the groin: but he continued to lead the grenadiers forward on the right while the 43rd and 28th wheeled to the left and cleared the coverts where the Canadians were holding up the 78th. At this moment a third bullet passed right through Wolfe's lungs, and he fell to the ground. Lieutenant Brown and two men carried him towards the rear hoping to find a surgeon, but he assured them it was needless, and sank at once into unconsciousness. The voice of victory recalled him for a moment. "They run, see how they run!" cried Lieutenant Brown. "Who runs?" asked Wolfe with great earnestness, "like a person roused from sleep." "The enemy, sir," replied Brown. "Egad, they give way everywhere." Wolfe had one last order to give. "Go, one of you, my lads, to Colonel Burton—tell him to march Webb's regiment (the 48th) with all speed down to Charles's river, to cut off the retreat of the fugitives from the bridge." Then, having finished his work, he turned on his side, saying only, "Now, God be praised, I will die in peace," and so expired.

By this time the morning drizzle had cleared away and the sun shone out: the Highlanders were far on towards the St. Charles river, and the 58th close to St. John's gate: on the right the grenadiers were also near the town, and guns were beginning to open on both of these advanced forces. Monckton had already fallen wounded near the left front of the 43rd: but Townshend hurried up, called off the pursuit, and ordered the whole line to dress and recover their former ground. He and Brigadier Murray then went to the head of every regiment, congratulated the officers, and thanked the men "for their extraordinary good be-

haviour." Rations were served out, entrenching tools, 1759
guns and ammunition were brought up: for the French still held the city and all their lines with a force of six or seven thousand men.

But the fight was over: Montcalm and his second in command were both dying of their wounds, and their army was rushing wildly across the St. Charles bridge. At nine o'clock that night Vaudreuil moved his headquarters thirty miles up the river and left Quebec to a garrison of less than 1500 men. Montcalm died before dawn. Three days later Admiral Saunders brought the fleet into the basin and trained his guns on the lower town: the army, too, had now 118 guns in position. The French sent a flag of truce and agreed to surrender the town "upon condition that it is not relieved before morning by the troops under MM. de Levis and de Bougainville." These commanders did not move. Accordingly, on the 18th of September the garrison marched out with the honours of war, and Quebec became a city of the British Empire.

CHAPTER II

1759–1764

Renewed French activity—The British besieged in Quebec—Defeated in battle at Sillery—Relieved by the Fleet—Expedition up the St. Lawrence—Concentration of three armies on Montreal—Fall of the French Empire in Canada—Capture of Martinique—Siege and capture of Havannah—The 43rd in England again.

1759 To take so watchful an enemy by surprise, to land so large a force in good order and to gain such formidable heights without loss—this was the brilliant part of Wolfe's achievement: the actual fighting was a comparatively simple matter. The battle was a contest of orthodox manœuvres and hard hitting: the French were the larger army, the British the better soldiers. As to this we have the word of Montcalm himself. "If I could survive this wound," he said as he lay dying, "I would engage to beat three times the amount of such forces as I commanded with a third their number of British troops." The casualty lists go some way towards confirming this view. The French army, out of 7520 men men engaged, lost over 1500: the British 664 out of 4829, only 9 officers and 50 men being killed.

The 43rd—no doubt from their central position and the accuracy of their own fire—suffered but slightly. Out of 327 men engaged only three were killed: Ensign Lewis Jones, two sergeants and eighteen

privates were wounded, and two missing. The regi- 1759
ment was taken into action by Major Elliott, Colonel James being in charge of the camp. On the 18th these two officers were both employed on important duties: Major Elliott with 500 men dislodged the enemy from a strong entrenchment on the north side of the St. Charles river, taking prisoners, guns and ammunition: and Colonel James, in command of detachments from the 43rd and other regiments, paid the last honours to their dead general. The body had been embalmed and was now to be taken to England: the troops marched in procession as far as the water side at Point Levi, where the 84-gun ship *Royal William* was lying off in readiness, and there they watched with every sign of grief while the great three-decker went from them down the river.

On September 23rd, being Sunday, the army attended a thanksgiving service on the field of battle, at which, Captain Knox records, an excellent sermon was preached. On the 24th, the troops were reviewed by General Monckton, who was leaving for New York on sick leave. With him went the Adjutant-General, Major Barré, who was also wounded: and Captain the Hon. Richard Maitland, of the 43rd, was appointed Deputy Adjutant-General in his place. On the 29th a return was made of the effective strength of the garrison: the grand total of the ten regiments, together with rangers and artillery, was 7313 of all ranks: of which the 43rd mustered 585 rank and file, with one lieut.-colonel, one major, seven captains, ten lieutenants, seven ensigns, one chaplain, one adjutant, one surgeon, one surgeon's mate, one quartermaster, twenty-nine sergeants, eighteen drummers and two fifes. The figures, interesting as they are, do not

1759 quite complete the picture of military life in 1759: the return of the women "allowed to be victualled" as being on the establishment of the several regiments was not made until February 1760. It enumerates 569 in all, of whom 63 belonged to the 43rd. They must have been fine strong women, fit to be the wives and mothers of soldiers. Captain Knox records that in the whole course of that severe winter not one of them died or was even sickly—on the contrary "the sergeant who brought me this return reported them all well, able to eat their allowance, and fit for duty both by day and night."

On October 18 the remainder of the fleet sailed for England, leaving with the army only two sloops, the *Racehorse* and the *Porcupine.* This encouraged the French, who were again becoming active. The army which M. de Levis had collected to the west of Quebec numbered about 18,000 in all—12,000 Canadians, 1200 Acadians, two troops of light infantry, thirty companies of marines and five battalions of veteran troops. Man for man they were not the equal of the British, but they were in better health, better acquainted with the country and much better supplied. Their first plan of attack was an escalade, to be carried out on a
1760 night towards the end of February. The men were exercised with snowshoes and scaling ladders, but the garrison got wind of their preparations and the attempt was never made. They did, however, cross to the south side of the river, and take possession of Point Levi: but from this post they were ejected by our light troops, supported by a detachment of the 43rd. An attempt to counter-attack a few days later was driven off by Murray, who held the position with the light troops and led four battalions over the ice to

take the assailants in the rear. The result was a hasty retreat with a loss of sixty killed, wounded and prisoners. 1760

But small successes of this kind did little to avert the real danger. The British garrison were holding Quebec under conditions immensely more difficult than those which Montcalm had failed to meet. They had a ruined city to defend with inadequate numbers, they had no regular communications open with the surrounding country, and the lack of fresh food was rapidly destroying their health. By March 1 the force of 7300 was reduced to 4800 fighting men, chiefly by fever, dysentery and scorbutic diseases, and this in spite of great care and skill on the part of the commanding officers and remarkable sobriety on the part of the troops. On one day—March 12—the 43rd lost a lieutenant and five privates: either their quarters were bad or they were overworked, for Captain Knox notes that "this is the most unhealthy corps in the garrison."

Worst of all, however, was the fear that M. de Levis was using Wolfe's methods for the taking of Quebec. By the middle of April, when the ice had begun to break up in a rapid thaw, it was reported that a force of 12,000 men was to be joined by a fleet of seven frigates and sloops under M. Vaugeulin, and with sixty days' preserved rations to hover about the city and search out the place to strike. An armed schooner was hastily equipped, and sailed on the 23rd from Orleans under command of a lieutenant of the 35th, who had been formerly in the navy: he was to make for Halifax and "hasten up our fleet, in case they have not yet sailed, by acquainting the Admiral or Commodore of our precarious situation."

1760 The French fleet did not appear, but on April 27 their upstream flotilla, consisting of several frigates and sloops and innumerable galiotes, floating batteries, and batteaus, dropped down to Cape Rouge: Levis and his army occupied the village of St. Foy, and his advanced posts came in touch with the light troops of the garrison. At seven o'clock on the morning of the 28th the British army marched out to the Heights of Abraham with entrenching tools, sixteen field guns and two howitzers. The French van was visible on the high ground towards Sillery, and the bulk of the army advancing on the right from St. Foy. On seeing them, General Murray, who had already formed his line, ordered his troops to throw down their entrenching tools and attack at once before the enemy could deploy. His front line was composed of the 48th, the 15th and the 2nd battalion of the 60th Royal Americans on the right, under Colonel Burton: the 28th, 78th, and 47th on the left, under Colonel Fraser; and the 58th, and 43rd as right and left centre, under Colonel James. In the second line were the 35th and the 3rd battalion of the 60th: the flanks were covered by light infantry and rangers, and the artillery were to take ground in front in the intervals, or on the flanks as circumstances might require.

The British attack was made briskly and was at first completely successful: the artillery did "amazing execution": the French van gave way, first on the wings and then in the centre. But in the meantime the main body of the army, massed in heavy columns, broke upon Murray's right flank, driving in the light infantry and pressing hard upon the right brigade of the line: at the same moment another division under Levis himself attacked the left flank. Murray brought

up the 35th and 60th to prolong his front to right and 1760
left, but the French, bent on cutting off the garrison from the city, persisted in their double flanking movement, and in this they had the great advantage of superior numbers. On the left of our line they took two redoubts, and though Colonel James by a gallant counter-attack with the 43rd and 60th recaptured these works and held them for a time, his losses were so heavy that he was at last compelled to retire.

The situation of the British troops was now desperate : in the rush of their first attack they had abandoned their own position on the heights and plunged into swampy ground, where they had to fight almost knee-deep in melting snow and water. The order was given to " fall back "—an unaccustomed and alarming order which was answered by shouts of " Damn it, what is falling back but retreating? " The retreat was accordingly begun : the guns were all abandoned and many of the wounded, but the army got away in good order and the pursuit was not pressed. The enemy had, in fact, been very roughly handled, and their best troops had suffered most. They contented themselves with entrenching on a line about half a mile from the city walls. The garrison at once opened on them with shot and shell, and probably cheered themselves by doing so. They needed encouragement, for they had lost in this battle of Sillery 1100 men out of 3140 engaged.

On the following day the French began siege operations : their boats landed stores, artillery and provisions at the Foulon, and their army threw up a line of counter-vallation with embrasures, platforms and batteries. But the British artillery was more than equal to theirs and inflicted losses upon them

1760 which eventually ran up to 500 killed and wounded. On May 1 the *Racehorse* sailed for Louisbourg and Halifax to urge the fleet once more to come to the relief of Quebec. But the dangerous moment had already passed: the garrison had worked indefatigably at their defences, and it was no longer possible to carry the place by storm, though in Captain Knox's belief this might have been done on the 28th, 29th, or 30th of April, "before the soldiers recollected themselves." They had now recovered their good humour and energy, and were completely organized: every regiment had its alarm post—that of the 43rd was "the stockades by the citadel next Cape Diamond," and that of their convalescents was "Vaudreuil House."

On the morning of May 9 the garrison had the "inconceivable satisfaction to behold the *Leostoffe* (*Lowstoffe*) frigate sail up into the basin." There was at first a doubt as to her nationality: but when Captain Deane saluted with twenty-four guns and put off in his barge, officers and soldiers alike "mounted the parapets in the face of the enemy and huzza'd with their hats in the air for almost an hour . . . the gunners were so elated that they did nothing but fire and load for a considerable time." In the evening the London newspapers, with the news of Hawke's great victory over Conflans in Quiberon Bay, were sent over to the French generals under a flag of truce. The parole and counter-sign were ordered to be Deane and Swanton—for Swanton in the *Vanguard* was reported to be following Deane. He arrived on the 15th, in company with the frigate *Diana* and the express schooner which had summoned them: there was again "unspeakable joy" in Quebec, and the gunners "immediately gave the enemy a general discharge of all our artillery, three times repeated 1760 without any return."

Early next morning the *Vanguard*, with the two frigates, went up with the flood tide and attacked the French squadron, which made only a poor defence. Commodore Swanton burned the *Pomona*, drove the *Atalanta* ashore and set her on fire, and took and destroyed all the rest except the *Marie*, a small sloop who threw her guns overboard and escaped to St. Peter's Lake. After "this notable morning's business" he fell down to Sillery, laid the *Vanguard's* broadside to the right flank of the enemy's trenches, and in a few hours made them entirely untenable.

The town batteries also played upon the trenches, but when deserters brought the news of a general French retreat they were ordered to fire with the guns elevated in order to overtake the enemy in their flight. For two hours the cannonade was tremendous—even at Bergen op Zoon, Knox says, there had been nothing like it, and the ground was afterwards found to have been ploughed up for two miles. The French guns made no reply, but their rearguard fired a single volley of musketry. After this a reconnoitring party found the trenches abandoned: and early on the 17th General Murray sent his light troops in pursuit, following himself with detachments from all his ten regiments. The enemy, however, got away with the loss of a few stragglers and all their wounded. Their total losses were now ascertained: in the battle of Sillery and the siege operations they had had 152 officers and over 2000 men killed and wounded.

On the 18th Lord Colville's fleet arrived before Quebec. On the 19th the *Hunter* sloop went upstream to recall the *Lowstoffe* and *Diana*: the latter returned

1760 with her, but the *Lowstoffe* ran upon some uncharted rocks and sank at once; the crew were taken off by the *Diana.* The loss was a light one at such a moment: the whole French squadron was destroyed and its commodore captured; the army was in flight and an immense booty of guns, ammunition and stores in the hands of the victors. The garrison had come triumphantly through a very trying siege: they had buried a thousand of their comrades, and had suffered so heavily from sickness that their effective strength was not more than 4000 men at the moment of greatest strain. Their success was due to something more than mere doggedness: Captain Knox attributes it to "the active examples and abilities of our Governors, together with the most exact discipline, observed and supported throughout by the officers of every rank: the great harmony and unanimity which has subsisted among the several corps, even to the private men, and between them and their superiors as one family: the unparalleled humanity to the sick and wounded, and the invariable attention displayed on every occasion to the preservation of the health of the soldiery." This is in all probability a just and well-founded statement: but it is almost more interesting as handing down to us the ideals of the 43rd a century and a half ago.

King's ships and traders now began to arrive daily at Quebec, and the position there being completely secure, the army's attention was turned in another direction. On June 3 came the news that General Amherst was moving on Montreal, and that Sir William Johnson was also on his way there with a large body of Indians. Vaudreuil, however, was defiant, and it was decided to send more troops from Quebec. Out of 5200 men in the garrison some 2500 were now fit 1760 for service, and the whole of these were ordered to prepare for the expedition. The *Porcupine* sloop, two armed vessels, eight floating batteries and twenty flat-bottomed boats were put under command of Captain Deane: the guns were got on board, and the troops were reviewed on July 12 by General Murray. They were formed into seven battalions; the detachments of the 43rd and the 2nd Royal Americans contributed the whole of the 4th battalion under Major Oswald, and the same two regiments with three others supplied a battalion of grenadiers. The total strength of the 43rd draft was thirteen officers, ten sergeants, ten corporals and 207 men: a surprising number when we consider that the regiment had only put 327 into the field at the taking of Quebec nearly ten months before.

The expedition embarked on the 13th and started upstream next day with a fair wind. On the 16th they reached the Rapids and began to work through them under fire from a French battery: two transports grounded but were got safely off under cover of the *Porcupine's* guns. Some days were then spent in landing parties to clear the country, and the Rapids were finally passed by the whole fleet on the 26th. On the 29th the flat-bottomed boats were sent back to bring up the Louisbourg division which was following: they rejoined on August 12, having put the troops ashore at Point Champlain. In the meantime French forces had made their appearance on the banks of the river and occasionally fired at the ships.

On August 16 the army went ashore on the island of Ignatius, that the transports might be cleaned out and aired. The Louisbourg division now appeared,

1760 far astern, and when they had joined next morning the whole fleet weighed together. L'Isle au Noix, St. John's and St. Teresa were successively occupied, and on September 4 the whole expedition disembarked on the north side of the island of St. Teresa. From there General Murray and Colonel Burton pushed on next morning to reinforce Brigadier Haviland, leaving Lord Rollo to command on the island. The same day there came the news that the Commander-in-Chief's army had reached Perrot Island, within less than four leagues of the city of Montreal.

The net was now drawing rapidly in : Murray and Burton returned to bring on their remaining troops, and on the 7th they landed without opposition at Pointe de Tremble on the island of Montreal, within eleven miles of the city : on the 8th, at noon, they encamped close under the Cape, or Mount, of Montreal, and General Murray took up his quarters in the suburbs. The three British armies were now all in position, but if they hoped for fighting they were disappointed. Vaudreuil had opened negotiations on the evening of the 7th, the capitulations were signed on the 8th, and Colonel Haldiman took possession of the city the same day. A brilliant feat of concentration, without a battle and almost without loss, had ended the French Empire in Canada.

The 43rd returned immediately to Quebec and went into cantonments for the winter. Their muster, taken at Quebec on October 2, 1760, shows a strength (on paper) of 27 officers, 30 sergeants, 30 corporals, 17 drummers, and 440 rank and file.

1761 In the early part of 1761 the regiment was selected to join in an expedition against the West Indian Islands. The force was to be commanded by General

Monckton, and it consisted of eleven battalions, 1761
besides rangers and artillery. The 43rd were at first encamped at Staten Island, where Lieut.-Colonel Dalling, lately commanding the light infantry, joined and took over the regiment in succession to Colonel James, who had retired. Major Robert Elliott, who had commanded it at the battle of the Heights of Abraham, had been appointed Lieut.-Colonel of the 55th.

The expedition sailed from New York on November 19: at Barbados they received drafts and regiments which brought the force up to 12,000 men, and on
January 7, 1762, they landed at Martinique and moved 1762
on Port Royal. Morne Tortenson and Morne Garnier, the two hills which dominated the town and citadel, were captured before the end of the month, and the surrender of the place followed. In the two fights the 43rd had two privates killed and three officers and six privates wounded. The capital, St. Pierre, was a strong place, but the French decided not to defend it. The whole island was surrendered to General Monckton, and the capitulation of Grenada, St. Lucia and St. Vincent followed.

War was now declared with Spain, and an expedition sailed from Portsmouth to attack the Spanish West Indies. A part of Monckton's force—including the 43rd—was ordered to join: they fell in with the fleet off Cape Nicholas on May 7, and brought the whole armament up to 10,000 men in 150 transports convoyed by 37 ships of war. A further force of 4000 was to come from New York. The chiefs of the expedition were Admiral Sir George Pocock and General Lord Albemarle.

The Admiral began at once to show the spirit of his service: instead of keeping safely to the south

1762 and rounding Cape St. Antonio, he ran along the north shore of Cuba and brought his immense convoy through the Straits of Bahama—a dangerous but justifiable saving of time. At Havannah he found the Spanish fleet lying in the famous harbour, protected by strong forts, a boom and three ships sunk in the fairway. The Spaniards would not fight, but neither could they prevent a landing: while the Admiral made a demonstration to the west, Commodore Keppel and Captain Harvey, with a detached squadron, threw the troops ashore on the east.

The army was divided into two corps, one of which advanced to Guarda Vacoa, while the other, in which the 43rd formed part of the 3rd brigade, was to attack the great fort known as the Moro, a bomb-proof work upon the summit of a high, steep rock. The labour and difficulty of this attack were immense: the guns had to be dragged over a rocky shore in such heat that several men dropped dead from exhaustion, sandbags had to be carried forward ready filled with earth, and the enemy's fire had to be kept down as far as possible. Finally, when the batteries were completed, and a sortie by the Spaniards beaten back, the fort proved too strong for the guns of both fleet and army, and the ships had to sheer off with considerable loss after a fierce engagement of some hours. Four days afterwards, on July 3, the besiegers' principal battery took fire and was totally destroyed. It had taken 600 men seventeen days to make it.

Discouragement and sickness followed: at one time there were 3000 seamen and 5000 soldiers down with malaria and other fevers, suffering too from want of water and fresh provisions. But the men were veterans, and not to be beaten: they made new 1762 batteries and worked them untiringly. By July 20 they had silenced the guns of the fort, beaten in the upper works and got a foothold in the covered way. On the 22nd arrived more siege material, and on the 28th the reinforcements from New York.

It was now decided to drive a mine beneath the defences: but at this point a new and formidable obstacle was discovered—a cutting in the solid rock, eighty feet deep and forty wide, lay between the outer approach and the walls of the fort. One narrow ridge of rock had been left to cover the end of the cutting from the sea front, and across this the miners succeeded in passing to the dead ground under the walls. The enemy made a desperate counter-attack, sending 1200 men across the harbour to climb the hills in our rear, but they were repulsed with great loss, and on July 30 the mine was sprung.

Only half the east bastion was destroyed, and the breach made was a narrow one. But the storming party advanced against it so boldly that the defenders fled. The Governor, Don Luis de Velasco, and his second in command, the Marquis de Gonzales, were of the old Spanish tradition: they gathered a hundred good men round the colours and refused all terms. The end soon came: Don Luis himself fell mortally wounded, and died after the Castilian fashion, courteously offering his sword to an enemy worthy of it.

The fort was immediately reorganized, and fresh batteries built to command the town. On August 10, the Governor having declined a summons from Lord Albemarle, fire was opened from all the British batteries with overwhelming effect. In six hours almost all the enemy's guns were silenced, and white flags

1762 appeared in every part of the defences. The Governor then surrendered upon the usual capitulations.

The British army's loss in this siege was considerable—eleven officers were killed, nineteen wounded, four died of wounds and thirty-nine of disease. Of the men 279 were killed, 631 wounded, 52 died of wounds and 657 of disease. The officers were severely tried, for their losses from disease were fifty per cent. greater in proportion than those of the rank and file. The 43rd were exceptionally fortunate in this respect: Captain Spendlove, wounded, was their only casualty among the officers, and they lost only thirteen men from fever, with ten killed in action, fifteen wounded and four missing. They had, however, nearly a hundred cases in hospital when they left for Jamaica.

Havannah was by far the richest city in the West Indies, and the prize money amounted to three millions sterling. It was allotted in the following proportions: to the Admiral and General Commanding-in-Chief, £122,697 10*s*. 6*d*. each; to the captains of ships, £1600 each; to field officers, £564 14*s*. 6*d*. each; to the naval lieutenants, £234 13*s*. 3*d*. each; to captains in the army, £184 4*s*. 7¼*d*. each; to sailors and marines, £3 14*s*. 9½*d*. each, and £4 1*s*. 8*d*. each to privates in the army.

1764 The 43rd remained at Jamaica till March 1764. On the 19th of that month they were relieved by the 36th, and sailed for England, reaching Portsmouth in July. On October 2 they were reviewed at Chatham by Major-General Parslow, who reported that they were, "For the numbers, a good regiment, well appointed and fit for service." The remarks appended on details are interesting: the uniform was red faced with white, and the officers' coats were laced with gold.

The officers were "armed with fusils, which they are desirous they may be permitted to make use of instead of espontons (or half-pikes)." The sergeants had halberds: the men were well clothed and accoutred, but their black guetres (gaiters) were "not well fitted: white bespoke." The drill was good, the evolutions being "performed according to the new regulations, and surprisingly well for the time they had to practice"; the fire discipline is marked "the whole according to order, and well." 1764

This report was earned in spite of the fact that the 43rd had come home much below strength only two months before and that the new recruits "appeared in the ranks." It is a remarkable proof of the efficiency of the regiment as a school for soldiers, and of the enduring influence of Wolfe as a commander of men.

American War of Independence

CHAPTER III

1765–1782

The American War of Independence—First war service of the 52nd—With the 43rd at Boston—Lexington—Bunker's Hill—Brooklyn—White Plains—Fort Washington—Rhode Island — Brandywine — Wayne's affair — Biggerstown — Return of the 52nd to England—The 43rd at the Surrender of New York.

1765 THE 52nd spent the first ten years of their regimental existence in England and Ireland. In June 1765 they embarked at Cork for their first foreign service, and in August of that year they landed at Quebec. Their uniform at this time was a red coat faced and lined with buff, and ornamented with white lace with a red worm and one orange stripe: a buff waistcoat and breeches and black gaiters.

1774 The regiment remained in Canada until 1774, when it was ordered to Boston to reinforce General Gage. They found there ten other regiments of infantry, of which the 43rd had been the first to land in June. The situation was critical. The resistance of the Colonists to taxation of imports had been put to the test by the arrival in December 1773 of Lord North's cargo of dutiable tea. The tea had been seized, as all the world knows, and thrown into Boston harbour: the British Parliament had retaliated by a Boston Port Bill; the Colonists had petitioned the King; the King had refused to receive their petition. In spite of all that Chatham and Burke could do, conciliation was defeated in both Houses, and coercion was decided upon. The Colonists then refused to supply British troops with barracks or clothing, and by February 1775 they were collecting guns and military stores for armed resistance. 1775

Gage's fateful move was made on April 18, 1775. On that night he sent a secret expedition to seize some stores reported at Concord, eighteen miles from Boston. The force was composed of some marines, with about 800 grenadiers and light infantry from several regiments, including both the 43rd and 52nd. They were ferried across the Charles river to East Cambridge, and six companies, under Major Pitcairn, were sent forward to secure a bridge over the Cambridge river.

Their movement had been seen: the bells of Boston gave the alarm and mounted messengers spread it. When the troops reached Lexington at four o'clock they saw through the early morning mist a body of seventy "minute men" drawn up on the village green. No one knows who "fired the shot heard round the world": the Colonists were summoned to disperse and shots came from both sides. Pitcairn's horse was wounded and one soldier; of their opponents seven were killed and nine wounded. The Colonists then gave way: a hundred men of the 43rd were left to guard the bridge, and by seven o'clock the remainder of the force had reached Concord.

After destroying the stores Colonel Smith began his return march. But by this time the whole countryside was up and snipers were firing from behind every kind of cover. The British troops began to run short of ammunition; they had in all nearly thirty-six miles

1775 to cover, and wounded to carry. When Lord Percy met them at Lexington with a relief force they were still marching and fighting gallantly, but "with their tongues hanging out of their mouths like dogs after a chase." The losses of the 43rd were Lieutenant Hull mortally wounded and a prisoner, four men killed, five wounded, and two missing. The 52nd lost three killed, two wounded and one missing—a small affair for their first touch of active service, but whatever they thought of it then, a fight as memorable as any in the history of nations.

Congress now assumed the title of "The United Colonies," and voted an army. The whole province of Massachusetts rose in arms; a force of 20,000 militia invested Boston on the land side, and they were soon reinforced by a large body of Connecticut men under Colonel Putnam. They were provided with artillery, and began, during the night of June 16, to entrench themselves on Bunker's Hill, a high piece of ground beyond the river. This move was discovered at daybreak on the 17th; the fleet at once opened fire on the new works, and a force was dispatched to land and drive the Colonists from them.

The troops detailed for this attack were ten companies of grenadiers, ten of light infantry, the 5th, 38th, 43rd and 52nd regiments, and some field artillery, under command of Major-General Howe and Brigadier-General Pigott. They were embarked from Boston at midday and landed on the opposite shore soon after three o'clock under cover of the guns of the fleet. They were then formed up with the light infantry on the right, the grenadiers on the left, the 5th and 38th in support, and the 43rd and 52nd as a third line. The fleet and field artillery began to bombard the enemy's lines, and the whole force moved uphill to 1775 the attack.

The Americans' position was a strong one: their right flank was covered by a large body of men concealed in the houses of Charleston; on their left was the new breastwork, while the ground in front was a steep ascent knee-deep in thick grass and intersected by walls and fences. Their fire was reserved until the British troops were within point-blank range, and was then delivered with deadly effect. Nearly the whole of the front rank went down before it, and the survivors, weighed down with heavy packs under a burning sun, failed to drive their attack home. Their officers called on them for a second effort, and they responded gallantly, but were again beaten off with heavy loss.

At this point General Clinton, who had been watching the fight from Copt's Hill, came over by boat as a volunteer with a small reinforcement, and a fresh attack was organized. General Howe had noted, during the first advance, what appeared to be a weak point in the lines between a breastwork and a rail fence. He decided accordingly to make a feint against the fence and to press home his real attack against the breastwork and redoubt. The men were ordered to use the bayonet only: they threw down their packs, and some even took off their coats; the General put himself at the head of the grenadiers on the left wing, and the artillery raked the breastwork, driving the defenders out of it into the redoubt.

This attack was successful, but not immediately. The troops in front of the redoubt were held up by some brick-kilns and enclosure walls and exposed to a concentrated fire; and when General Howe had

1775 forced his way into the redoubt from the left his men were driven back and he was for a moment or two alone among the enemy. At last the bayonet was victorious, and the Colonists were driven out. They retreated over Charleston Neck, and though they made three counter-attacks during the night they were repulsed by the advanced post held by the remains of the 52nd.

The British casualties were very heavy—19 officers and 207 men killed, 70 officers and 758 men wounded. The 43rd had 12 men killed, and 4 officers and 82 men wounded. The 52nd had 3 officers and 21 men killed, and 8 officers and 80 men wounded. Heavy as the losses of both regiments were for a fight on so small a scale, they were not quite a full proportion of the whole. This was probably due to the fact that during the first and most murderous fire neither regiment was in the first line; during the final attack they suffered perhaps more heavily than the other corps.

Bunker's Hill was not only an expensive victory, but an unfruitful one. The victors kept and fortified the position they had taken, but they were still blockaded and had now two garrisons to maintain instead of one. They had to face an enemy of great force and spirit; they were closely shut in and suffering from lack of fresh provisions and from the heat of the summer; depression fell upon them and sickness soon followed.

In July General Washington appeared before Boston. He had been appointed Commander-in-Chief of the American forces and was rapidly raising troops from all classes, even from the Quakers, who perceived then, as Quakers have perceived in a more recent crisis, that in the last resort freedom and security rest upon force. It is believed that at one time Washington had not less than 200,000 men under arms or in training, and their enthusiasm, their knowledge of the country and their possession of its resources, made them a truly formidable army. They had no regiments to match ours, but with a sufficient superiority in numbers, mobility and supply, a militia may defeat the best of regulars. 1775

As the weather changed from summer to winter the position of the British forces in Boston became more anxious still. The arrival of a corps of light cavalry from Ireland increased the crowding and diminished the available supplies. Live cattle, vegetables, firewood and coal, had to be shipped over from England. They were sent in large quantities, but some of the ships bringing them were lost at sea: others were captured by the enemy at the harbour mouth. Snow and cold winds set in—an additional hardship for men who had to sleep in tents and do without fuel.

The officers worked hard and directed even their amusements to keeping up the spirits of the garrison. As in some of our Arctic expeditions and in our prisoners' camps of to-day, the stage was a great resource. The following is an entry in the Journal of Lieutenant Martin Hunter, of the 52nd. "During the winter, plays were acted at Boston twice a week by the officers and some ladies. A farce called *The Blockade of Boston,* written by General Burgoyne, was acted. The enemy knew the night it was to be performed, and made an attack on the hill at Charleston at the very hour the farce began: they fired some shots and surprised and carried off a sergeant's guard. We immediately turned

1776 out and manned the works, and a shot being fired by one of our advanced sentries, a firing commenced at the redoubt, and could not be stopped for some time. An orderly sergeant standing outside the playhouse door, who heard the firing, immediately running into the playhouse, got upon the stage, crying out, 'Turn out! turn out! they're hard at it, hammer and tongs.' The whole audience, supposing the sergeant was acting a part in the farce, loudly applauded, and there was such a noise he could not for some time make himself heard. When the applause was over he again cried out, 'What the d—— are ye all about? If ye won't believe me, be Jasus ye need only go to the door, and there ye'll hear and see both.' If the enemy intended to stop the farce they certainly succeeded, as the officers immediately left the playhouse and joined their regiments."

The winter passed without serious fighting, but the initiative lay clearly with the Colonists. Their indignation had been fired afresh by the arrival of a copy of the King's Speech, rejecting their petition, and they had now changed their red flag, which was merely revolutionary, for one of thirteen red and white stripes, intended to symbolize the union of the thirteen colonies and destined to represent a new and powerful nation. They had also prepared an offensive stroke which took the British entirely by surprise.

In the beginning of March 1776, when General Gage had gone to England on leave, a fortified battery suddenly made its appearance on the Heights of Dorchester, which commanded the town and part of the harbour. This was the work of a strong force under General Thomas; they had carried out Washington's orders with such secrecy and energy that Howe

declared they had done more in one night than his 1776
whole army could have accomplished in months. They had done even more than he knew at first: for on a closer inspection he found that they had forestalled his intended attack by a still stronger work behind.

The battery opened fire and at once drove the ships of war from their moorings: the town, too, was clearly becoming untenable. On the 8th a flag of truce was sent to General Washington through the "select men" of Boston, informing him that General Howe would leave the town undamaged if he were allowed to withdraw unmolested. Washington stood out for a more direct proposal; whereupon Howe declined to treat, embarked his men and baggage and sailed for Halifax, taking with him 2000 American Loyalists. Washington made no attempt to interfere, but marched into the town with his drums beating and colours flying, while the British troops were going on board.

The 43rd and 52nd, with other regiments, reached Halifax on the 4th of April and stayed there until June, when they sailed again with the expedition under General Howe for Staten Island. They disembarked on July 3, a critical moment, for on the following day the American Congress issued their Declaration of Independence and all hopes of a peaceful issue were laid aside.

Reinforcements, including a body of 13,000 subsidized Hessians and Brunswickers, had already been voted by the Home Parliament; the first division of them had now arrived, and the army was regularly formed. The 43rd were placed under Brigadier-General Smith in the 5th brigade, the 52nd under Major-General

1776 Valentine Jones in the 3rd brigade, of Lord Percy's division. Admiral Lord Howe arrived with his fleet a few days afterwards, and after a fruitless attempt to obtain an interview with Washington he decided to attack General Putnam, who was entrenched with a strong force at Brooklyn. On August 22 the British troops, covered by the fleet, landed without opposition on the south-west end of Long Island, and the enemy's outpost detachments were withdrawn to the wooded hills which covered Brooklyn. On the night of the 26th a pass in these hills was seized, and on the following day General Howe's attack was delivered. The Americans fought well, but suffered a complete disaster: they were driven from their position with a loss of 3000 men, and were greatly discouraged by their failure. The British loss was very slight: it amounted to only 350 in all, the 43rd having one man killed and two officers wounded, the 52nd one officer and one man killed, and one officer and eight men wounded or missing. Among the killed was also Lieutenant Doyley, of the Guards, who was attached to the 52nd.

On September 15 General Clinton followed up this blow by taking a large force up the river and landing at Keff's Bay, three miles north of New York. The Americans thereupon abandoned their artillery and stores, and retreated to the north end of the island. The city was occupied and garrisoned by the British, but within a few days a third part of it was burnt to the ground by incendiaries.

In the second week of October the 43rd and 52nd again embarked with Howe's river expedition: they landed at Pell's Point and took part in a smart skirmish on the 18th. On the 28th they were both engaged in

the action at White Plains, where Howe defeated 1776
Washington with very small loss. On November 16 they both took part in the attack on Fort Washington, one of the two forts which commanded the Hudson river. Lieutenant Martin Hunter, of the 52nd, relates in his journal a curious and probably unique incident which occurred here. "The light infantry," he says, "embarked at King's Bridge in flat-bottomed boats and proceeded up the East river under a very heavy cannonade. They landed and stormed a battery, and afterwards took possession of a hill that commanded the fort [Washington]. Before landing, the fire of cannon and musquetry was so heavy, that the sailors quitted their oars and lay down in the bottom of the boats; and had not the soldiers taken the oars and pulled on shore, we must have remained in this situation. The instant we landed the enemy retreated to Fort Washington, and on our carrying the outworks it was summoned and surrendered." The charge which "carried the outworks" appears to have been a fine one: and it resulted in the surrender of 2700 of the Americans, in addition to 170 taken during the same day by the 42nd Highlanders. On December 1 Lord Percy and General Clinton sailed from New York to attack Rhode Island, the principal base of the American privateers. The expedition landed at daybreak on the 8th, at Weaver's Bay on the west of the island, and forced the enemy to retire to the mainland with the loss of some prisoners and two guns. Newport, the capital town, was occupied, and for the remainder of the winter the 52nd were cantoned on the island.

In the following year, 1777, the battalion companies 1777
of the 43rd were also on Rhode Island: the flank

1777 companies formed part of Lord Cornwallis' force which attempted the relief of Trenton on the Delaware. In June the 52nd left for Staten Island, and on August 22 they assisted in repelling an American attack made by a force under General Sullivan. In September the battalion companies of the regiment formed part of a small army which invaded Jersey under General Clinton.

In the same month the flank companies of both regiments marched with General Howe against Washington, who had advanced to Brandywine Creek in order to oppose a British move on Philadelphia. The Americans met the attack of the light infantry and the German chasseurs with a heavy fire of musketry and artillery, but when the Guards and grenadiers advanced from the right they fell back into the woods and were pursued closely for two miles. After an ineffectual stand near Dilworth they retired on Chester and Philadelphia, darkness falling in time to save them from annihilation. They owned to a loss of 300 killed, 600 wounded, 400 prisoners and eleven guns captured: among the wounded was the French volunteer, the Marquis de la Fayette. The British casualties were 90 killed and 450 wounded. After this defeat the enemy abandoned Philadelphia.

On September 20 Sir William Howe received information that General Wayne with 1500 men and four guns were lying concealed in some woods a few miles off. He promptly sent Major-General Sir Charles Grey (afterwards Earl Grey) to surprise this force by a night march, with two regiments and a composite battalion of light infantry. Lieutenant Martin Hunter gives the following account of the affair: "As soon as it was dark the whole battalion got under arms,

Major-General Grey then came up to the battalion 1777
and told Major Maitland, who commanded, that the battalion was going on a night expedition to try and surprise a camp, and that if any men were loaded they must immediately draw their pieces. The Major said the whole of the battalion were always loaded, and that if he would only allow them to remain so, he [the Major] would be answerable that they did not fire a shot. The General then said if he could place that dependence on the battalion they should remain loaded, but that firing might be attended with very serious consequences. We remained loaded, and marched at eight in the evening to surprise General Wayne's camp.

"We did not meet a patrol or vidette of the enemy till within a mile or two of the camp, when our advanced guard was challenged by two videttes. They challenged twice, fired, and galloped off full speed. A little further on there was a blacksmith's forge: a party was immediately sent to bring the blacksmith, and he informed us that the piquet was only a few hundred yards up the road. He was ordered to conduct us to the camp: and we had not marched a quarter of a mile when the piquet challenged, fired a volley, and retreated. General Grey then came to the head of the battalion and cried out, 'Dash on, Light Infantry!' and without saying a word the whole battalion dashed into the wood, and guided by the straggling fire of the piquet, that was followed close up, we entered the camp, and gave such a cheer as made the wood echo.

"The enemy were completely surprised, some with arms and others without, running in all directions in the utmost confusion. The light infantry bayonetted

1777 every man they came up with. The camp was immediately set on fire, and this and the cries of the wounded formed altogether one of the most dreadful scenes I ever beheld. Every man that fired was instantly put to death. Captain Wolfe was killed, and I received a shot in my right hand soon after we entered the camp. I saw the fellow present at me, and was running up to him when he fired. He was immediately killed. The enemy were pursued for two miles. I kept up till I got faint from loss of blood, and was obliged to sit down. Wayne's Brigade was to have marched at one in the morning to attack our battalion while crossing the Schuylkill river, and we surprised them at twelve. Four hundred and sixty of the enemy were counted next morning lying dead, and not one shot was fired by us—all was done with the bayonet. We had only twenty killed and wounded."

This ruthless fight was not soon forgotten by the Americans, who were the equals of professional soldiers in dash but not in stoicism. A fortnight after the night surprise, Washington threw his whole force suddenly against the light infantry battalion at Biggerstown, an advanced post a mile in front of Germantown, north of Philadelphia. Lieutenant Martin Hunter's narrative is as follows: "General Wayne commanded the advance and fully expected to be revenged for the surprise we had given him. When the first shots were fired at our piquets, so much had we all Wayne's affair in remembrance, that the battalion were out and under arms in a minute. At this time the day had just broke, but it was a very foggy morning, and so dark we could not see a hundred yards before us. Just as the battalion had formed, the piquets came in and said the enemy were advancing in force. They had hardly joined the battalion when 1777 we heard a loud cry of 'Have at the blood hounds! revenge Wayne's affair!' and they immediately fired a volley. We gave them one in return, cheered and charged. As it was near the end of the campaign our battalion was very weak: it did not consist of more than 300 men, and we had no support nearer than Germantown, a mile in our rear. On our charging they gave way on all sides, but again and again renewed the attack with fresh troops and greater force. We charged them twice, till the battalion was so reduced by killed and wounded that the bugle was sounded to retreat; indeed, had we not retreated at the very time we did, we should all have been taken or killed, as two columns of the enemy had nearly got round our flank. But this was the first time we had ever retreated from the Americans, and it was with great difficulty we could get the men to obey our orders.

"The enemy were kept so long in check that the two brigades (our supports) had advanced to the entrance of Biggerstown when they met our battalion retreating. By this time General Howe had come up, and seeing the battalion retreating all broken, he got into a passion and exclaimed, 'For shame, Light Infantry! I never saw you retreat before; form! form! it's only a scouting party.' However, he was soon convinced it was more than a scouting party, as the heads of the enemy's columns soon appeared. One, coming through Biggerstown with three pieces of cannon in their front, immediately fired with grape at the crowd that was standing with General Howe under a large chestnut tree. I think I never saw people enjoy a discharge of grape before, but we really

1777 all felt pleased to see the enemy make such an appearance, and to hear the grape rattle about the Commander-in-Chief's ears, after he had accused the battalion of having run away from a scouting party. He rode off immediately full speed, and we joined the two brigades that were now formed a little way in our rear; but it was not possible for them to make any stand against Washington's whole army, and they all retreated to Germantown except Colonel Musgrave, who with the 40th regiment nobly defended Tewes House until we were reinforced from Philadelphia."

The battle ended badly for Washington: General Grey brought up the left wing of the army, and after a stiff fight completely broke the Americans. Lord Cornwallis came up at this moment: his light horse joined in the pursuit, but three battalions of grenadiers, including those of the 43rd and 52nd, had run themselves out of breath in a vain attempt to reach the battlefield before the end. Our total loss was only 535, but among the seventy killed were Brigadier-General Agnew and Lieut.-Colonel Bird. The Americans lost nearly 1300, including General North killed and 54 officers captured.

The winter passed without movement on either
1778 side. In May 1778 Sir William Howe was succeeded as Commander-in-Chief by Sir Henry Clinton; and as the King of France had now concluded a treaty with the Americans by which he agreed to give them active assistance, it was decided to concentrate the army at New York. The northern force in Canada, under Generals Burgoyne and Carlton, had been driven to surrender in the previous October.

In June, therefore, Philadelphia was evacuated

and the army, with an ample provision train twelve 1778
miles long, moved north. On the 28th, as the last division descended the hills above Freehold in New Jersey, the Americans made an attempt to cut off the rearguard and baggage column. They were smartly repulsed by the grenadiers and Guards, who lost very few men from the enemy's fire, but fifty-nine from sunstroke. The Grenadier Company of the 52nd had Lieutenant Francis Grose wounded, and their captain, John Powell, killed. This was the fourth captain of grenadiers the 52nd had lost during the war, and the drummer of the company remarked, not unnaturally, though no doubt mistakenly, "Well, I wonder who they'll get to accept of our grenadier company now: I'll be damned if I would!"

The army reached Neversink on the 30th and embarked for New York. For the 52nd the last act of the campaign was the surprise and capture of Lady Washington's dragoons by two battalions of light infantry. "While at New Bridge," says Lieutenant Hunter, "we heard of their being within twenty-five miles of our camp, and a plan was laid to surprise them. We set out after dark, mounted behind dragoons, and so perfectly secure did the enemy think themselves that not even a sentry was posted. Not a shot was fired, and the whole regiment of dragoons, except a few who were bayonetted, were taken prisoners."

The 52nd was now so greatly below strength that it was ordered home. Those of the men who were fit for service and volunteered to remain were drafted into other corps: the regiment itself, mustering ninety-two effective men, reached England in December 1778.

A harder lot fell to the 43rd. They were at

1778 Quaker's Hill on September 29, where they helped to beat off General Sullivan's attack on Newport; the Commander-in-Chief in his general orders particularly distinguished "with great applause the spirited exertions of the 43rd under Colonel March." They then endured for three successive campaigns the weariness of garrison and other operations in an unsuccessful
1781 war. In May 1781 they joined the field force in Virginia, and after beating the Pennsylvanians and La Fayette's Continentals at James Town, and marching more than 1100 miles to Williamsburg, they were, with the rest of Cornwallis' army, driven into York Town and shut up there on the 30th of August 1781.

A fleet of twenty-six of the line, with 7000 troops, was dispatched to relieve them, but not in time. The enemy opened their parallels on October 6, and began the bombardment on the 9th, with forty guns and sixteen mortars of eight and sixteen inches. On the fourteenth they stormed two redoubts, and in spite of a very gallant sortie by 250 men under Lieut.-Colonel Abercrombie they completed their second parallel next day. Cornwallis had now only one hundred shells left and hardly a gun mounted; a violent storm prevented an attempt to cross the river at midnight; at daybreak on the 19th the batteries opened on the town, and the British force of 5000 men surrendered to an army of 13,000 Americans and 9000 French. The same day the relieving army, under Sir Henry Clinton, sailed from New York after a fortnight's delay.

The strength of the 43rd at the surrender amounted to 94 rank and file, with 168 sick and wounded. With the rest of Cornwallis' army they had won the admiration of their enemies, and especially of the French,

who have always valued rightly the soldierly virtues. 1782
From the Americans, too, they received a treatment that British prisoners of war have often desired in vain: they were well clothed, supplied with necessaries, allowed the same rations as the American armies, encamped as far as possible by regiments, and an inspecting officer was allotted to every fifty men. In November 1782, within fourteen months of their capture, peace was signed at Paris, and early in 1783 they returned to England.

CHAPTER IV

1782–1799

Regiments and counties in 1782—The 52nd in the East Indies—War with Tippoo Sahib—The storming of Cannanore—The Rock Fort of Savendroog—With Cornwallis at Seringapatam—The battle in the Dark—The 43rd in the West Indies—At Martinique again—The death-trap of Guadaloupe—The surrender at Point Bacchus.

1782 IN 1782 his Majesty sent to each regiment of the line an intimation of his pleasure that county titles should be conferred on the infantry, in order that a connection between the regiments and the counties might be cultivated, to facilitate the procuring of recruits. The 43rd thereupon received the title of THE MONMOUTHSHIRE REGIMENT, and the 52nd that of THE OXFORDSHIRE REGIMENT.

1783 In 1783 troops were required in the East Indies for the war against Tippoo Sahib, Sultan of Mysore. The 68th and 77th, who were under orders to embark, having successfully claimed their discharge according to the terms of their enlistment, on February 13 the 52nd were invited to engage for three years' Indian service at a bounty of three guineas apiece. In twenty-four hours the regiment was completed to one thousand men from other corps at Chatham, and it embarked at the beginning of March. The transports reached Madras in August, and there one of them, the *Kingston*,

blew up with a loss of 63 lives. She had on board, 1783
among other passengers, two hundred men, women and children belonging to the 52nd.

The first service of the regiment in India was the employment of one wing in a punitive expedition against Cannanore, on the Malabar coast, the Queen of which was an ally of Tippoo Sahib. The force was commanded by Brigadier-General Norman M'Leod, and his Brigade-Major was Captain Martin Hunter, of the 52nd. The journal of the latter gives some interesting details of the taking of Cannanore on December 14, 1783: "We marched after sunset, lay on our arms all night, and next morning made a move close to the principal fort. In taking possession of some commanding ground, the light infantry were attacked by four or five hundred of the enemy, armed with matchlocks, shields and swords: they only fired one volley and retired to the fort, but the light infantry were so much exposed in the attack that we had three officers and twenty-five men killed and wounded. It was doubtful if the fort could be stormed in the event of a breach being made, as we were uncertain of the depth of the ditch, and whether it was wet or dry. The General wished to ascertain these points before the battery opened. I had a letter from the Adjutant-General offering a large sum to any man of the battalion that would undertake this hazardous service. I read the letter to the men, and a man named Rowlandson Taylor, of the 52nd, who was an old American light infantryman, undertook and executed it so coolly and well, that he not only ascertained the exact depth of the ditch, but observed that it was wet except at the very point where we intended to breach it, and returned under a heavy fire of musquetry without being touched.

1783 General M'Leod was so much pleased that he gave him fifty guineas.

"Colonel Frederick applied to General M'Leod to have the honour of storming the breach with the Bombay Grenadiers, but was told he intended that honour for Captain Hunter and the light infantry. Two days after, the breach was thought practicable, and I received orders to hold the battalion in readiness to storm at one o'clock in the afternoon. Lieutenant Robinson commanded the 'forlorn hope,' consisting of a sergeant, corporal, and thirty volunteers from the battalion. At eleven o'clock the battalion paraded three companies in front: the men each carried a scaling ladder, and the remainder of the brigade fascines to fill up the ditch. We were supported by the Battalion Companies of the 6th and 52nd Regiments, and as one o'clock struck we advanced in close column to the breach, which was most gallantly defended, and carried after an obstinate resistance. Lieutenant Robinson and the 'forlorn hope' were nearly all killed or wounded, and in the battalion altogether 4 officers and 53 men."

The enemy made a stand behind the breach, and the 6th and 52nd turned aside to attack another small fort, where they were at first repulsed for lack of scaling ladders. Captain Hunter, after about half an hour, got his men away from plundering the first fort, and came on in support: the two regiments, seeing them, made another effort and carried the second fort just as they came up. The Queen was taken prisoner, and next day the town and the three remaining forts were surrendered.

Peace with Tippoo Sahib was concluded on March
1784 11, 1784, and the 52nd were ordered to the coast of

Coromandel, where they were cantoned for the greater 1790
part of the next six years. In March 1790, war having again been declared against Tippoo Sahib, the regiment was ordered to proceed to Trichinopoly under command of Captain Martin Hunter. After a march of twenty days they arrived at Trichinopoly on April 29th, and joined the army collecting there under the command of Major-General Musgrave. Coimbatoor was taken without difficulty on July 22, and on August 15 Dindigull was invested by a detached force consisting of the 52nd and some native troops. A night attack on the 21st was unsuccessful, owing to the nature of the ground, which was too rocky to be climbed in darkness : but the place was carried next morning with a loss to the 52nd of only 4 killed and 16 wounded. On September 10 Paulighautcherry was invested, and on the 22nd the garrison surrendered as the British troops were moving out of the trenches to the assault.

Two days later Tippoo's army was threatening Coimbatoor, and the detachment was recalled from Paulighautcherry to the main army. The rest of the year was spent in marching and manœuvring, without any success in bringing the enemy to action. On
January 29 Lord Cornwallis, the Governor-General and 1791
Commander-in-Chief, arrived from Bengal to take command, and after more marching and skirmishing the army sat down before the fort of Bangalore on March 5, 1791. On the 15th the batteries opened fire, and on the 21st, the breach being reported practicable, the assault was ordered. The storming party was composed of the grenadiers of the 36th, 52nd, 71st, 72nd, 74th and 76th Regiments, followed by their light companies, and led by a forlorn hope of 30 men, under Lieutenant James Duncan of the 71st and Lieu-

1791 tenant John Evans of the 52nd. The fighting was severe, but by one a.m. the fortress was taken. The losses of the 52nd during this siege were two officers and four men killed, and two officers and three men wounded. Lieutenant Evans, who led the forlorn hope, was among the wounded.

The Nizam of Hyderabad had now sent his cavalry, 15,000 in number, to co-operate with the British army. The two forces joined hands on April 13, and on May 4 they marched together against Seringapatam, Tippoo Sahib's capital. On the 13th they found the enemy in position on the river Cauvery, eight miles below the city. A night attack on the 14th failed from the badness of the roads and the weather, but on the following day, in spite of the extreme fatigue of the troops, an action was forced and the position taken. The enemy crossed the river and took refuge in the island upon which Seringapatam stands, where he had batteries entrenched.

The British army at once marched upstream to Canambaddy, some miles above the capital: but Lord Cornwallis, seeing that the weather had broken and the monsoons were upon him, decided that the siege must be postponed. He buried a great part of his ammunition and stores, destroyed his battering-train—which had mostly been captured from the enemy—and retired on Bangalore. On the march he was joined by the Mahratta army of 32,000 men, with 30 guns, under Hurry Punt and Purseram Bhow, and the Nizam's cavalry, being no longer fit for service, were allowed to return home. The rest of the year was spent in reducing three important forts. Nundydroog was taken on December 9, Savendroog on the 21st, and Outredroog on the 24th. The storming of Savendroog was a really remarkable feat, and the 52nd had a full share in the honour of it. The fortress was built on the side of a mountain, among almost inaccessible rocks, and the first difficulty was to find and secure a position for the siege batteries. During the night of December 10 the grenadiers of the 52nd and 72nd, and a battalion company of each, with the 26th Native Infantry in support, left the camps on this vitally important enterprise. They crossed a steep hill, let themselves down into a ravine by clutching the branches of trees growing on the side of the precipice, then ascended three hundred feet of rock by hand-over-hand climbing, and finally gained a terrace which overlooked the whole fortress from only 300 yards' distance. 1791

In ten days' time batteries had been built and breaches made in the walls: a daylight assault was ordered for the morning of the 21st. The storming party was under command of Lieut.-Colonel Colebrook Nesbitt, of the 52nd, and was directed to make four simultaneous attacks: the grenadiers of the 52nd and the flank companies of the 76th against the eastern hill to the left, the light company of the 52nd against the western hill on the right, the flank companies of the 71st against any works or forces that might be discovered between the hills, and other detached parties to work round to the enemy's rear: the main body of the 52nd and 72nd to follow their flank companies. The enemy, already shaken by the bombardment, could not face this array. When the 52nd advanced with their bands playing "Britons, strike home!" the grenadiers and light infantry went in with a rush and stormed one of the strongest forts in India with a loss of only five men wounded. Lord Cornwallis in the General Order could only attribute the collapse of the defenders "to their astonishment at seeing the good order and determined countenance with which the troops who were employed in the assault entered the breaches, and ascended precipices that have hitherto been considered in this country as inaccessible."

The time had now come for the attack on Seringa-
1792 patam. The army marched on February 1, 1792, and came in touch three days later with the enemy's cavalry. Tippoo Sahib himself lay with his main army in a strongly entrenched camp, completely covering the city, which lies on a triangular island in the Cauvery. This island was surrounded by redoubts, and the north-west angle of it was occupied by the great fortress of Seringapatam, only separated from the camp by a fordable branch of the river. In order to besiege the fortress effectually it was necessary first to drive the mobile army from the camp and also to take the out-lying redoubts upon the north front of the island itself.

The entrenchments were reconnoitred on February 6, and as soon as it was dark the attack was organized in three columns: on the right the 36th and 76th, under General Medows; in the centre the 52nd, 71st, and 74th, under the Commander-in-Chief himself; and on the left the 72nd Highlanders, under Lieut.-Colonel Maxwell, of the 74th. The native troops were divided among the three columns. The whole force moved off at eight o'clock, and after marching five miles in pitchy darkness the centre and right columns came up against the abattis in front of the camp. The left column skirted a line of hills two miles to the east, but upon striking the enemy's camp on its extreme right flank, converged upon the centre and crossed the river at the same point as the troops of the left

centre. General Medows in the meantime was also 1792
bringing the right column over to the left, passing along the whole front of the position, so that eventually, when the enemy fell back upon the island and fortress of Seringapatam, the whole British force came into the attack delivered by the centre column.

Probably no battle was ever so confused as this. It was fought in darkness, against an enemy greatly superior in numbers, who, after flying in apparent panic, rallied at every possible opportunity and counter-attacked with disconcerting noise and fury. The first clash came near the right centre of Tippoo's line, where the advanced companies of Cornwallis' column broke through between two small forts known as the Sultan's Redoubt and Sybald's Redoubt. They drove the enemy across the river, mixed with them in pell-mell pursuit, and landed close to the eastern face of the fortress. Captain Lindsay actually collected the grenadiers of the 71st upon the glacis, and would have attempted to rush the place, but the bridge was raised a moment too soon. He was then joined by the grenadiers of the 76th and some of the light infantry of the 52nd, and with them forced his way right down to the Llal Baugh, or "Garden of Pearls," in the extreme east corner of the island, where he was surrounded and had to cut his way through with the bayonet. After this, he fell in successively with the grenadiers of the 74th, with Lieut.-Colonel Knox of the 36th (formerly of the 43rd) and Lieut.-Colonel Baird, leading the grenadiers of the 52nd, the light company of the 71st, and some of the 72nd from the left column. These detachments all forced their way across the island with the greatest courage and intelligence, but they were unable to clear the native town of the enemy, and returned at day-

1792 break towards the north bank of the river, where they could hear the sound of heavy fighting.

Meanwhile the battalion companies of the 52nd and 71st, with the 72nd from the left column, and some active troops, had also forded the river, driving another body of the enemy before them. They also found themselves close to the glacis of the fort, and took up their position in the Daulet Baugh, immediately to the east of it. What followed is well described by Captain Martin Hunter, commanding the 52nd: "The night was so dark, I did not know I was within range of the guns of Seringapatam. Tippoo soon found us out, and brought every gun he could to bear upon us, which determined me to recross the Cauvery, and try to join Lord Cornwallis, who I knew had halted somewhere near the Sultan's Redoubt with a part of the 71st Regiment and a battalion of Sepoys. Lord Cornwallis did not know that the 52nd was within less than a quarter of a mile of him, till within half an hour of the attack of Tippoo, who had recrossed the Cauvery with his whole force. The night was so dark, the first intimation we had of their approach was from the 'tom-toms,' followed by cheering and a volley. They were within 200 yards of us when the Regiment was ordered to fire a volley and to charge. In this charge I was dangerously wounded, and carried into the Sultan's Redoubt: the Regiment thought I was killed.

"Lord Cornwallis had fallen back with his small bodyguard, and sent orders to the 52nd to retreat, which orders were delivered to Captain Conran, next in command of the Regiment. At this time the men were under a galling fire from the enemy, and, getting impatient, they called out in the hearing of Captain Conran: 'Had Captain Hunter been alive, he would

have ordered another charge at those black rascals !' 1792
Conran said, 'Well, my lads, *though I have received orders to retreat*, you shall have another dash at them !' This charge, in my opinion, was the saving of Lord Cornwallis and the few troops he had with him—the 52nd covering his retreat till he got beyond the Baugh Bridge, when Tippoo gave up the pursuit and bent his whole force against Sybald's Redoubt. Had not the 52nd recrossed the Cauvery, and by the greatest good luck fallen in with Lord Cornwallis, he must inevitably have been taken by Tippoo."

This account is confirmed by Major Dirom, who says that Tippoo's counter-attack was made about two hours before daylight, and was beaten off by Lord Cornwallis' troops with the help of the seven battalion companies of the 52nd and three companies of the 14th Bengal Native Infantry, who had just time to replace their wet cartridges after recrossing the river, when Tippoo's centre and left rallied and came down upon them. "It was near daylight," he says, "before they were finally repulsed."

After a battle like this, as full of brilliant episodes as a book of the *Iliad*, every regiment will have its own story to tell, and the whole will be impossible to reconstruct. It is enough for us that every corps that night illuminated the darkness with its own ardour and constancy. Lord Cornwallis recorded in his General Order next day that the zeal and gallantry displayed "was universal through all ranks to a degree that has rarely been equalled." The result was in doubt till daylight came, but not for a moment afterwards. The British army was encamped within less than two miles of the fortress, 80 of Tippoo's guns were in their hands, and 20,000 of his men were killed or wounded. Siege

1792 works were pushed forward rapidly, but no second blow was necessary. In less than three weeks articles of peace had been signed, and Tippoo had delivered his two young sons as hostages: he paid for his cruelty and treachery by the loss of half his dominions and a large indemnity, and no penalty was ever more thoroughly deserved.

In a battle with barbarous hordes a smaller civilized force has to face the chance of annihilation: but the alternative is victory at small cost. The British losses at Seringapatam were comparatively slight: even the 52nd had only one officer and ten men killed, and five officers and thirty-six men wounded or missing. After a month's rest they were able to begin the long march back to cantonments: and by October they were at Poonamallee, near Madras.

1793 In 1793, the National Convention of France having declared war against England, it was thought desirable to secure the small settlement of Pondicherry, the only remaining (and still remaining) fragment of that Franco-Indian Empire for which Dupleix and Suffren had fought so stoutly. The 52nd were employed in the siege, which ended on August 22 in the surrender of the French, and the regiment then returned to Poonamallee. Two years later it was again actively engaged in the expedition against the Dutch settlements in Ceylon, and after a completely successful campaign of eighteen months returned to India, and was stationed at Tanjore, in the Carnatic. In the following year, 1797, the regiment was ordered home, and the privates fit for further service were drafted into the 77th and 80th with a bounty of three guineas apiece. On February
1798 19, 1798, the 52nd embarked for England after a service of fifteen years in India.

The 43rd were also on active service during the 1793
earlier part of the fresh struggle with France. On November 17, 1793, they were embarked for the West Indies, under command of Lieut.-Colonel Drummond. They reached Barbados at the end of January, and sailed again a few days afterwards for Martinique, with Sir Charles Grey's squadron. The army consisted of ten regiments of infantry, three battalions of grenadiers, three of light infantry, and a detachment of dragoons. It was divided into three brigades, the 43rd being in the third brigade, commanded nominally by H.R.H. Prince Edward, but until his arrival from Canada by Lieut.-Colonel Sir Charles Gordon.

At first all went well with the expedition. On February 8 a landing was forced by Gordon's division at Cas Navire—the same spot where the 43rd had landed for the same purpose thirty-two years before—and four days later the regiment, with five companies of grenadiers, distinguished itself in a turning movement by crossing four ravines and capturing all the batteries which defended them. The brigade then advanced to within a league of Fort Bourbon.

The second brigade, under General Dundas, landed on the 17th in the Bay of St. Pierre. As in 1762, the French did not attempt to defend their capital, but concentrated on Fort Bourbon, and after a short siege surrendered there on March 25. The 43rd had one casualty during the attack on this fort, which is probably unique in the list of personal misfortunes. Captain Burnet, while leading his men, was blown up by an explosion which blackened him and set his clothes on fire. He was saved by those nearest to him, but was then wounded by a bullet which broke his arm. After this, his own grenadiers mistook him for a French

1793 black, and bayoneted him in several places. Happily for everyone concerned, he survived all his injuries, and, though wounded again four months afterwards, he lived to command the 17th and retire in 1805.

A garrison of five regiments under General Prescott was left in Martinique : the remainder of the force was embarked by Sir Charles Grey for St. Lucia, which surrendered at once. Two regiments were left there, and the squadron proceeded to Guadaloupe. A detachment, which included one company of the 43rd, landed under a heavy cannonade and carried Fort Fleur d'Epée with the bayonet alone. The remainder of the brigade was then disembarked, and the whole island fell into their hands. But yellow fever and other diseases soon reduced so small a garrison to a condition of extreme weakness. In June a French naval and military expedition of 2000 men succeeded in landing, and at the third attempt they carried Fort Fleur d'Epée by storm. Lieut.-Colonel Drummond, with the 43rd and some British merchants and sailors who had volunteered to help him, succeeded in making good their retreat, but all attempts to expel the French forces from the island were in vain : the climate was more deadly than the enemy. The strength of the 43rd on September 1 was twenty-three rank and file fit for duty, and 176 sick.

On September 26 the French drove Colonel Drummond and his convalescents out of Petit Bourg, took the hospital, and killed all the sick. Colonel Drummond was then surrounded at Point Bacchus and compelled to surrender. The same fate overtook Brigadier-General Graham and his force at Berville : another detachment at Fort Matilde succeeded in re-embarking. Of the 43rd, five captains, three lieutenants, one ensign,

two quartermasters and one surgeon had died during 1795
the campaign. Lieut.-Colonel Drummond, three captains, four lieutenants, two ensigns, and the other surgeon were prisoners on a hulk at Pointe à Pitre. Early in 1795 they succeeded in seizing a boat and escaping, and the remaining prisoners were exchanged not long afterwards. The regiment—for ten officers and a few privates were still the 43rd Regiment—then returned to England and was stationed at Monmouth to recruit. Two years afterwards the 43rd, a thousand strong, under the command of Colonel Drummond, again sailed for Martinique, where they garrisoned Fort George. In 1798 they were ordered to St. Pierre, where they remained two years, suffering terribly from
the climate. In February 1800 they returned, only 1800
300 strong, to Fort George, and on April 25 they sailed for England—a striking example of the truth that a regiment may die many deaths and yet remain immortally efficient.

Sir John Moore and the 52nd

CHAPTER V

1799–1808

Sir John Moore Colonel of the 52nd—His new system—The 52nd becomes a Light Infantry Regiment—The 43rd follows—The brigade at Shorncliffe—The expedition to Copenhagen, 1807—To Sweden, 1808—Moore's army sails for the Peninsula.

HITHERTO the service of the 43rd and 52nd had been like that of many other fine regiments, brilliant but not unique: they had proved their quality among the best. The time was now approaching when they were both to be drawn within the sphere of a new influence and launched upon a new career: in a few years to raise themselves from excellency to immortality, and to gain not merely a place of fame, but a place apart in the history of their country. "Six years of warfare," said Napier, "could not detect a flaw in their system, nor were they ever matched in courage or skill. These three regiments (the Rifles were the third) were avowedly the best that England ever had under arms."

The threads out of which this singular destiny was woven may be traced briefly as follows. In December
1799 1799 the King was pleased to order that a second battalion should be added to the 52nd, with Major-General Moore as Colonel-Commandant. This was done by enlisting two thousand volunteers from the militia: and in the following June the two battalions

SIR JOHN MOORE, COLONEL OF THE FIFTY-SECOND.

were embarked for service at Quiberon, and afterwards 1800
on the coast of Spain. In the attack near Ferrol on August 26th they were both engaged, and the first battalion "had the principal share in this action." They re-embarked for Lisbon next day, and returned to England in January 1801. On May 8 Major-General 1801
John Moore, Colonel-Commandant, was appointed Colonel of the Regiment in succession to General Cyrus Trapaud, deceased.

On January 10, 1803, his Majesty signified his 1803
pleasure that the 2nd battalion of the 52nd should be numbered the 96th Regiment of Foot, and on the 18th he further ordered that "the 1st battalion, which will then become the entire 52nd Regiment, shall be formed into a corps of *Light Infantry*, retaining, however, its present number and distinction of the Oxfordshire Regiment of Foot, and in every respect its rank in the service." On July 9 the 52nd Light Infantry and the 95th Rifles, with three other regiments, were formed into a brigade at Shorncliffe under command of General Moore. In reconstituting the 52nd he had been allowed as Colonel of the Regiment to select from the 2nd battalion such men as he judged to be the best adapted for the Light Infantry, and to replace them from the 1st battalion by men less calculated for such service. At the present day, when all regular infantry are trained under the same regulations and employed in the same kind of service, such an order would have no meaning. In the eighteenth century the light companies were composed of the most active and intelligent men in the regiment, and when a special force was needed to lead an important attack, the light infantry or grenadiers of a number of regiments were formed into composite battalions for

1803 the occasion. Such battalions possessed every element of efficiency but one—they had not the perfect cohesion and spirit of comradeship which belong to a true regiment. It became clear that if a fighting force must have a proportion of special infantry, then an army must have regiments of such infantry: Wellington went further—he formed a whole division of it, a *corps d'élite* upon which he could rely absolutely in the first or last resort. In our day we have not this specialization: it would cause serious difficulties in action, and the great advance in the average of efficiency has made it unnecessary. We still have better and worse among our regiments, but in the armies of 1914 and 1915 the *corps d'élite* have far outnumbered the rest: and this they have done by extending the tradition of the old Light Infantry.

It is suggested (in the Royal Military Calendar for 1820) that the new training adopted in 1803 was not originally invented by Moore himself—that "the improved system of marching, platoon-exercise, and drill, were entirely Lieut.-Colonel Mackenzie's." The words quoted may well be true, but they relate to matters of detail. Moore's ideas were more general: they went beyond any mere system of marching or exercise. According to his brother and biographer, James Carrick Moore, they were first formulated in a conversation with the Duke of York, who, as Commander-in-Chief, frequently consulted General Moore. "The Duke mentioned to him a design of enlisting some regiments of riflemen, a species of troops which had never been raised in this country. On which Moore observed, that our army was not so numerous as to admit of having enough of those for each detached force, which the nature of our warfare required. He therefore advised that some good regiments should be practised as marksmen, with the usual muskets, and instructed both in light infantry manœuvres, and also to act, when required, as a firm battalion. His Royal Highness approved this idea, and requested him to form his own regiment on that plan: and as many of the men were unfit for these complex duties, he was empowered to exchange them for more powerful and active soldiers, selected from another battalion. He then commenced this new discipline, and in a short time formed a regiment which for celerity and expertness was admired by experienced officers." 1803

The writer of these words clearly knew the facts, and from his account we can reconstruct with certainty the "new model" of Moore's imagination. The ideal Light Infantry regiment was to differ from others in being composed of "more powerful and active soldiers" fitted for "complex duties." They were to be sharpshooters, their manœuvring was to show "celerity and expertness," and at the same time they were to have all the order and cohesion of a "firm battalion." This combination of qualities is expected now of every good regiment: a hundred years ago it seemed almost unnatural. Even the best officers were not all so accomplished then: many had a belief in rigid formations as the only reliable ones. Robert Blakeney relates a typical conversation between General Graham and Colonel Browne at Barrosa. Browne was ordered to attack in open order, and was not pleased. A moment afterwards the General rode back, saying, "I must show something more serious than skirmishing: close the men into compact battalion." "That I will with pleasure," cried the Colonel, "for it is more in my way than light bobbing." In fact, the men—a

1803 composite corps of light infantry—were more versatile than their commander.

Moore's energy from the first swept aside all prejudices and difficulties: the time and the place, too, were in his favour. The country about Shorncliffe, as Captain Moorsom of the 52nd remarks in his *Historical Record*, was well adapted for Light Infantry manœuvres, "and at this period the threatened invasion was peculiarly favourable to the formation of a light corps, as every individual was kept in the same constant state of activity and vigilance as if absolutely in presence of an enemy, and the careful superintendence of Major-General Moore infused a soul and spirit throughout every rank which made them perform their various duties with a zeal and alacrity seldom attained in other corps." He adds that the regiment was accustomed to parade in light service order, and that Moore himself detailed very minutely what portion of necessaries each soldier was to carry: the heavy baggage was put into store at Gravesend, and the officers were only permitted to retain in camp a small portmanteau and their beds. The whole brigade was expected to be formed in column, with baggage packed and tents struck, ready to move off in one hour after receiving the preparatory order for march.

When Moore began forming his new model the 43rd were stationed in Guernsey, but by what seems a kind of natural affinity they soon followed the 52nd into the sphere of his influence. By an order of July 17, 1803, they were similarly transformed into a Light
1804 Infantry regiment; in January 1804 they returned to England, and in June they arrived at Shorncliffe and became part of the brigade which already contained the 52nd and 95th. The union was marked, two

SIR WILLIAM F. P. NAPIER, K.C.B., LT.-COLONEL OF THE 43RD.

months later, by an interchange which, however little 1804
noticed at the time, now seems to bear a symbolic meaning, like the transmission of blood between more primitive warriors: a company in the 43rd was given to a lieutenant of the 52nd. This young officer was William Patrick Napier, a soldier whose long and brilliant service in war was destined to be lost in the splendour of his fame as a historian.

On August 23rd the brigade at Shorncliffe was reviewed by the Commander-in-Chief, and on the following day the 52nd manœuvred alone before him. His admiration of the regiment's performance was expressed in a letter dated August 29th, in which General Moore is informed that "his Royal Highness has been pleased to recommend to the King that the promotion should be more extensive in that corps than has been usually granted." This promotion, the General remarks in his Regimental Orders of the 31st, "exceeds perhaps whatever at any one period has been accorded to a regiment." In September of this year the King further signified his approval by conferring the Order of the Bath on General Moore.

The hopes set upon the 52nd and 43rd are also traceable in the order that both regiments should be increased by the addition of a second battalion. That of the 52nd dated from August 8, that of the 43rd from November 25, 1804. The first battalions of the two regiments were again reviewed together by
the Commander-in-Chief in the following August, and 1805
on September 4 they both received orders to hold themselves in readiness for immediate embarkation. These orders were countermanded within a few days, but they mark an intention to use the 43rd and 52nd as a single corps rather than as separate units.

1806 This intention took effect in the following year, when the expedition to Copenhagen was decided upon. The first battalion of the 52nd had sailed a few months before for Sicily, and were not yet returned: but the 2nd battalion was now brigaded with the 1st battalion of the 43rd, the 95th Rifles and the 92nd Highlanders, under the command of Major-General Sir Arthur Wellesley. On August 16 the brigade landed on the island of Zealand, eight miles north of Copenhagen, and was smartly engaged on the 26th, near the village of Kioge, with a superior force of Danish troops, who fled before a charge led by the 92nd, with the 43rd and 52nd in support. Ten guns were taken, with sixty officers and 1500 men.

Copenhagen was then summoned, but the terms being rejected, on September 2 the British batteries opened fire, and after a bombardment of six days the inevitable surrender followed. The expedition was a painful political necessity, but probably no one now doubts that it was a necessity. Bonaparte had just declared a blockade of the British Isles: the Russian and Swedish fleets had been placed at his service, and the two Emperors counted on securing the Danish fleet as well. The danger of such an addition to the strength of our enemies may be judged from the amount of the captures at Copenhagen: sixteen ships of the line, fifteen frigates, six brigs of war, twenty-five gunboats, a number of vessels on the stocks, nearly twenty-five hundred guns, and naval stores and equipment sufficient to fill ninety-two transports.

The troops were re-embarked on October 20, and the fleet, with its immense convoy, set sail for England in magnificent array. But on the last night of the

voyage a violent squall scattered some of the ships, 1806
and the *Syren* transport, with nearly 700 of the 43rd on board, was driven upon a sandbank. In the midst of the confusion which followed, the tranquillizing strains of the "Dead March in Saul" were heard. They came from the flute of Ensign Richard Henry Neale, a young officer who, after distinguishing himself on this occasion and at the battles of Vimiero and Corunna, sold out and took Holy Orders. The troops landed safely next day: the 43rd were ordered to Yarmouth, and the 52nd to the barracks at Deal.

On April 30, 1808, the 1st battalion of the 52nd 1808
were embarked at Ramsgate as part of the army which was being sent under Sir John Moore to defend Sweden against a combined attack by Russia, France and Denmark. The expedition reached Gottenburg on May 17 and lay in the roads until July 3, when it sailed again for England, the eccentric King of Sweden having proposed only impracticable schemes and attempted to detain the English general by force. By July 21 the fleet was anchored at Spithead, where the transports were revictualled, and ten days later they weighed again for Lisbon. By an order received from the Horse Guards before sailing the men's queues, or pigtails, were to be cut off and the custom of soaping the hair discontinued, and this made the voyage a memorable one for the rank and file of the army. Captain John Dobbs, then an ensign in the 52nd, says: "We landed near Vimiero in a heavy surf, with only the clothes we wore, a blanket, and a few days' provisions in our haversacks: we had no change of clothes till we arrived in Lisbon, for our baggage had gone on thither by sea: we used to wash our shirts in the nearest stream, and sit by, watching till they were dry: but the men had great

1808 joy, for they were *relieved from their hair-tying,* which was an operation grievous to be borne."

From the narrative of Robert Blakeney, then a lieutenant in the 28th Regiment, we learn that Sir John Moore's force had first begun to disembark at Figueira on August 19, but on hearing there that the French had been defeated at Roriça and that a second engagement was expected, the General hastily re-embarked and sailed on to Peniche, nearer to the scene of action. There was an almost dead calm, and as the transports crept slowly along the booming of cannon from the hills round Vimiero was distinctly audible. There was great excitement among the troops and much fretting at the delay: the war that was beginning could not fail to be a serious one, and they were missing the opening scene. Happily the 2nd battalions of both the 43rd and the 52nd were already there, as will appear presently.

CHAPTER VI

1808–1809

The Peninsular War—The 2nd battalions of the 43rd and 52nd with Sir Arthur Wellesley—The battle of Vimiero—Sir John Moore Commander-in-Chief—The 1st battalions of the 43rd and 52nd join the Army—The Great Retreat—The battle of Corunna—The death of Sir John Moore.

THE intervention of England in the Peninsula was the 1807
grasping of a great opportunity. Napoleon, after the failure of his northern coalition against the sea power of this country, turned to the south, and thought to find fresh resources for making himself master of Spain, Portugal, and the South American possessions of both. In October 1807 he agreed with the party of Ferdinand, the Spanish heir apparent, that France and Spain should divide between them the kingdom of Portugal, England's oldest ally. An invasion of 25,000 French troops under Junot was immediately successful: the Portuguese Royal Family fled overseas to Brazil. Napoleon then turned on his own ally. By his convention with Spain he had the right to send troops to Bayonne and to reinforce Junot: in December he secured the main road to Madrid with two corps of 53,000 men, while another of 12,000 occupied Barcelona.
On March 18, 1808, after a riot in the capital, Charles 1808
the Fourth abdicated: on the 23rd Murat entered Madrid as Commander-in-Chief of the French army in Spain: by May both Charles and his son Ferdinand

1808 had been enticed to Bayonne, made prisoners and forced to resign the crown. On June 15 Murat, now styled Lieutenant-General of Spain, presided at an Assembly of Notables, which accepted Joseph Bonaparte as king under a constitution presented to them by Napoleon.

Joseph entered Madrid on July 20, and was proclaimed on the 24th.

Europe looked on with stupefaction: but the Spaniards, united by sudden indignation, rose like one man. The news of their revolt was welcomed by all parties in England: "never had so happy an opportunity existed for Britain to strike a bold stroke for the rescue of the world." The blow was struck and failed: but it was repeated with characteristic courage and tenacity, and in seven years the rescue of the world was accomplished.

The first English force to land in Portugal was one of 9000 men under Sir Arthur Wellesley. He was joined at Mondego Bay by General Spencer with 4000 men, and moved north along the coast road to meet reinforcements from England. On August 17 he fought a successful action at Roriça, and advanced to Vimiero, where three days later he was joined by Anstruther's and Acland's brigades: the former of them containing the 2nd battalions of both the 43rd and the 52nd. The same night Junot was reported to be advancing with 20,000 men. He had, in fact, only 14,000.

The French general intended to make a surprise attack at daybreak, but the British position was a difficult one to reconnoitre, and he did not discover it until nine o'clock. He then perceived Anstruther's and Fane's brigades on a hill in front of Vimiero, and

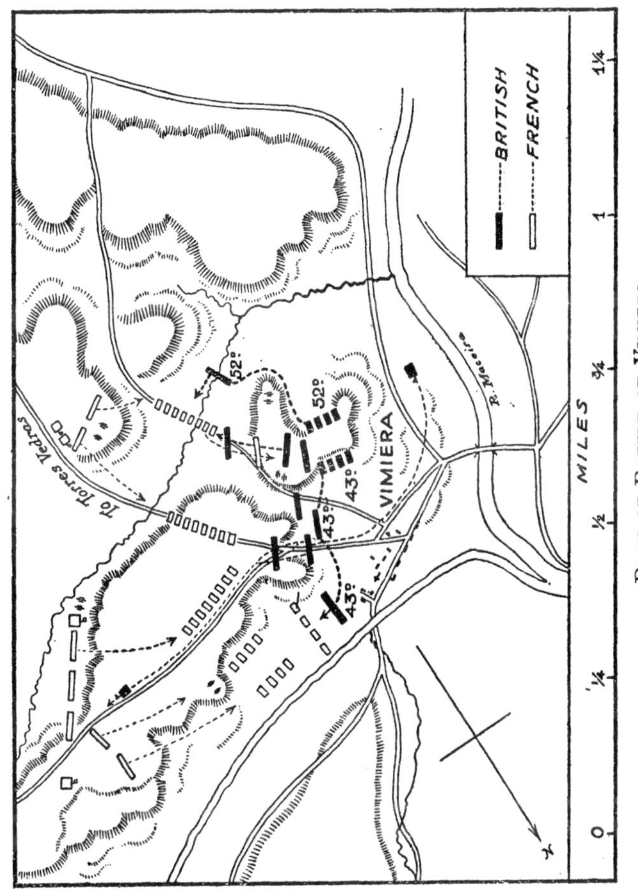

PLAN OF BATTLE OF VIMIERO.

1808 on a longer ridge, at an angle to their left, two brigades in an apparently detached position, two more connecting brigades being out of sight behind part of the ridge. He at once ordered Laborde and Brennier to attack simultaneously the two forces which he could see.

The attack went wrong from the first. Brennier found the English left separated from him by an unexpected ravine, and got entangled for some time among rocks and water-courses. Laborde engaged Fane and Anstruther, but the artillery of the 8th brigade, which was just then topping the ridge where the gap was imagined to be, halted and enfiladed him with deadly effect. Junot threw in Loison's corps on Laborde's left, and a triple attack was delivered on the two brigades of the British right. First a portion of Loison's troops drove in the skirmishers of the 60th and 95th on Anstruther's right, and inflicted heavy losses on a company of the 95th and three of the 52nd which covered their retirement. But on arriving on the front of the brigade this force was met by a charge of the 97th and at the same time taken in flank by the 52nd, who pursued as far as the nearest wood. Many of the English troops were mixed up with the flying enemy and in danger of being overlapped: but the 52nd, with admirable discipline, halted and covered their retirement from their too advanced position.

Immediately after this Laborde's main column attempted to rush Fane on Anstruther's left: they got within half-pistol shot, but there halted for a moment, out of breath from their climb, and were decimated by Colonel Robe's reserve artillery. At that moment the 50th, followed by the 9th and the 71st,

charged them with a yell, broke them with the bayonet, 1808
and scattered them in all directions.

The third attack was made by Kellermann with the reserve—a picked corps of Grenadiers and Swiss troops, who were intended to force their way past Fane's extreme left to the village of Vimiero. To assist in meeting this attempt Anstruther sent the 43rd across Fane's rear to occupy the churchyard. They had barely got there when Kellermann's Grenadiers came on at the double, carrying with them the remains of Laborde's corps and turning the left flank of the 50th, whose charge had taken them some way forward. A desperate fight took place in some vineyards, and the leading companies of the 43rd were beaten back into the narrower part of the ravine. The French grenadiers, following in a dense column, were taken in flank by the artillery of the 4th and 8th brigades, and the chance came for a decisive counter-stroke. It came to the right men. "Then," says Napier, "when the narrowness of the way and the sweep of the round shot was crushing and disordering the French ranks, the 43rd, rallying in one mass, went furiously down upon the very head of the column and with a short but fierce struggle drove it back in confusion. In this fight the British regiment suffered severely, and so close was the combat that Patrick, segeant-armourer of the 43rd, and a French soldier were found dead, still grasping their muskets with the bayonets driven through each body from breast to back!"

In the meantime the attack on the British left had been even less successful: Solignac had been wounded and repulsed with the loss of six guns: Brennier, after retaking the guns, had lost them again and was him-

1808 self wounded and captured. By pressing on with the five brigades of his left wing Sir Arthur Wellesley intended to cut Junot's army off from Lisbon and to isolate and capture Solignac and the rest of the artillery. But he was superseded at this moment by Sir Harry Burrard, who had just arrived from England, and who, after watching Sir Arthur win his victory, forbade him to use it. To make things doubly absurd, Burrard himself was superseded next day by Sir Hew Dalrymple: and while the unfortunate British army was thus passing from hand to hand by the mismanagement of politicians, their beaten opponents made good their retreat on Lisbon, marching the fifty miles at a stretch with only a two hours' halt. Their loss was thirteen guns, over 2000 killed and wounded, and a general and several hundred men prisoners. The British loss was much less than this, but one-sixth of it was borne by the 43rd alone: the regiment had 6 officers wounded, 40 men killed and 73 wounded, of whom no less than 48 afterwards died.

Junot had no desire to continue the contest: he sent in proposals at once, and on August 30 the convention of Cintra was signed, by which he agreed to evacuate Portugal, the British fleet to convey his army to France. In England there was much indignation at these terms, and Dalrymple was at once recalled: the other two Commanders-in-Chief had already gone home, and Sir John Moore was ordered to undertake the campaign against the French in Spain. Lieut.-General Sir David Baird was at the same time sent out with fresh forces, in which was included the 1st battalion of the 43rd, with orders to land at Corunna and effect a junction with Moore.

The advanced corps of the British army, in which

was Anstruther's brigade—the 95th and 20th regiments and the 1st battalion of the 52nd—left Lisbon on October 26, and reached Salamanca on November 21. The second battalions of the 43rd and 52nd were brigaded with the 9th, under Major-General William Carr Beresford, and marched as flank brigade with Sir John Moore's own army. Sir David Baird's force came in at Mayorga on December 20, and three days later the reserve—Anstruther's and Disney's brigades—marched upon Carrion to clear the way for a general attack on Soult at Soldanha. 1808

They were only a few miles forward on the deep snow-clogged road, when at midnight Captain George Thomas Napier, of the 52nd, overtook them with astonishing orders. They were to countermarch immediately, and by daylight next morning they were actually back again in the quarters they had just left. The reason of this was the arrival at Headquarters of a courier, bringing news that Napoleon had left Madrid with a very large army to cut Moore's line of communication with Lisbon. With Soult only a day's march ahead of him and Ney not much further off, Moore was in danger of being hemmed in by forces nearly three times as great as his own: the only course open to him was to retreat quickly through the north of Portugal and send for the transports to meet him. His danger was no new discovery to him: he had written from Salamanca a fortnight before: "The moment is a critical one: my own situation is particularly so: I have never seen it otherwise. But I have pushed into Spain at all hazards: this was the order of my Government, and it was the will of the people of England." He had to do what he could: his design was to draw away Napoleon from Madrid, do what

1808 damage was possible, and make his own escape in time.

The first part of the retreat was a race between Moore and Napoleon for Benavente, the junction of the two main roads of northern Portugal. Of these roads one runs through Astorga to Corunna, the other gives access, through Orense, to Vigo, where the transports were originally ordered to be in waiting. The leading columns, under Baird and Hope, marched at once: the reserve, with the Light Brigade and the Cavalry, followed next day and arrived at Mayorga late at night. On the 26th they marched to Castro Gonzalo on the Esla, and the Light Brigade, with two guns, was ordered to hold the passage of the river till the cavalry arrived. Here two privates of the 43rd, John Walton and Richard Jackson, greatly distinguished themselves. They were posted beyond the bridge, with orders that in case of attack one should stand firm while the other fired and ran back to the brow and gave notice whether the enemy were few or many. Towards night some chasseurs of the Imperial Guard made a dash to surprise the post. "Jackson fired, but was overtaken and received twelve or thirteen sabre cuts in an instant: nevertheless he came staggering on and gave the signal, while Walton, with equal resolution, stood his ground and wounded several of his assailants, who retired leaving him unhurt, but with his cap, knapsack, belts and musket cut in above twenty places, his bayonet bent double, bloody to the hilt, and notched like a saw." Napier notes this as "a remarkable display of courage and discipline"—in later days it would have earned a cross. Jackson recovered of his wounds, and lived to return in safety to England.

The Light Brigade spent the next two days and 1808 nights in destroying the bridge at Castro Gonzalo, two companies of the 43rd, under Captains Napier and Lloyd, working unrelieved, while the rest of the battalion helped to keep off the enemy. They rejoined the army at Benavente and were near being burned to death the same night in the convent where they were quartered. The place was densely packed with 6000 men and horses: a fire broke out and was enveloping a large window shutter, when Captain Lloyd, "a man of prodigious activity, strength and presence of mind, signalled to his companions to maintain silence, and springing on the nearest horse, he ran along the backs of the others, reached the blazing shutter, tore it from its hinges and threw it out of the window."

Benavente was not a comfortable billet: the inhabitants were unfriendly to the British, hid all their provisions and refused to supply the troops with food or drink, even when full and immediate payment was offered to them. When the 52nd arrived, drenched and cold, not a drop of wine could be found to revive them until Lieutenant Love discovered a recently bricked up wall in a convent outhouse. In spite of the abbot's denials and protests Love and two of his men let themselves down through a skylight, pulled the newly built wall to pieces, and found an inner chamber containing a huge vat of wine. While the wine was being carefully measured out to the soldiers, the abbot appeared through the trap-door and asked for a last drink before it was all gone. One of the men promptly seized him and plunged him head foremost into the vat, exclaiming: "By Jove, when the wine was *his* he was damned stingy about it: but now that it

1808 is *ours* we will show him what British hospitality is, and give him his fill" : and the abbot was only rescued from drowning in his own wine by the intervention of Lieutenant Love and his fellow officers.

On the 28th, after destroying all superfluous stores, the army left Benavente : the reserve followed next day. On the 31st the Light Brigade, under Brigadier-General Craufurd, was ordered to branch off to the left and make a cross-cut to the Vigo road : the main body marched on towards Corunna, and orders were sent to the transports to sail round and meet them there. With Craufurd went not only the 1st battalion of the 43rd, but also the 2nd battalion of the 52nd, which had been transferred to his command : and these two now drop out of the story of the retreat, for they reached their port and embarked for England without further difficulty.

1809 For the main body of the army the remainder of the march was a nightmare, a fantastic shifting scene of bravery, devotion, insubordination and debauchery. The trouble began at Astorga, when a corps of half-mad and half-starved Spaniards, under the Marquis of Romana, began pillaging. Their evil example entirely upset the weary and disappointed British troops, who spent the night in prowling about the town in search of wine and in quarrelling over billets. Even the personal interference of the Commander-in-Chief was unavailing to check their excesses : and this was only the beginning of the disgrace.

On entering Bembibre the Reserve found the whole place in the utmost disorder and confusion, and were kept at work all day in turning drunken stragglers out of the houses and trying to send them after their own divisions. Robert Blakeney, an eyewitness, says that the state in which the town had been left 1809 by the troops can only faintly be imagined. Every door and window was broken open in the insane search for wine: the huge vats had been pierced with bayonets, and the wine was running to waste, while round them, on the floor, lay the helpless and stupefied men. These wretches were collected and herded into the church, as the only building capable of holding so large a number. Those who succeeded in hiding themselves were overtaken next day by the French cavalry, who ruthlessly cut them down as they staggered along the road.

This extraordinary break-down of discipline is not difficult to explain: for most armies, certainly for a British army of that period, nothing could be so demoralizing as to be in continual retreat without any apparent cause. Insubordination was common among those who were furthest from the enemy, rare among those who were allowed to face him day by day. The Reserve, who had all the fighting, did hardly any of the drinking and plundering: and if they ever wavered it needed only a sight of the enemy to steady them. Their commanding officer, Sir Edward Paget, gave them a lesson at Calcabellos, where a few men, including at least one of the 52nd, were reported for plundering a deserted house in the town.

Next morning, January 3, the Reserve were marched out of Calcabellos and halted on the slope of a low hill. Here General Paget formed them into a hollow square, in the centre of which he held a drumhead court-martial on the plunderers. He continued the flogging of these men in spite of frequent reports of the enemy's approach up the further side of the hill. Finally of three culprits—Lewis of the 52nd, an

1809 Artilleryman and a Guardsman—two were fastened with ropes round their necks to the branches of a tree and held up on the shoulders of two men till the order to let them hang should be given. Just then a cavalry officer rode up to say his pickets were retiring. General Paget sent him angrily back to his men. Then " My God," he said, " is it not lamentable to think that, instead of preparing the troops confided to my command to receive the enemies of their country, I am preparing to hang two robbers? But though *that* angle of the square should be attacked I shall execute these villains in *this* angle." In the silence that followed this emphatic speech the hoofs of the retiring pickets could be clearly heard coming up the hill. " If I spare these three men," said the General, " will you promise to reform? " There was a breathless silence in the square, and he repeated his question. " If I spare the lives of these men, will you give me your word of honour as soldiers that you will reform? " For a moment there was not a sound, and then the whole square shouted " Yes." The culprits were released amid cheers, and at that instant the pickets, closely followed by the enemy's advance guard, appeared over the top of the hill.

The infantry were hastily withdrawn across the little river Guia, while General Paget held a strong position on the side of the hill and fired on the French cavalry as they came into sight on their way down to the river. In spite of this they pressed forward, charged furiously across the bridge, and did not turn back till they received the rifle fire of the 52nd and 95th, who were lining the hedges at the bottom of the hill. Great numbers were killed in this ill-advised attack, including their brave leader, General Colbert.

Sir John Moore had by now arrived from Villa 1809
Franca, and he ordered the 52nd further up the hill, for he saw that as soon as the French were across the river that regiment would be menaced on both flanks. The enemy crossed at first to the right of the position, and a good deal of confused fighting took place. A strong column also moved towards the left and attempted to cross the stream there, but our guns again opened such a destructive fire that they had to fall back, and the retreat became general.

By then it was quite dark, and under cover of night the Reserve withdrew and marched eighteen miles into Herrerias without leaving behind a single straggler—such was the moral effect of being allowed to look the enemy in the face.

But as they marched to Villa Franca on the way to Herrerias many of the Reserve, in spite of recent punishments and strict orders, could not resist sticking their bayonets into the salt meat that was being burnt in piles in the streets. In this way they carried off a substantial supper, which was shared that night even by some of the officers who had tried in vain to keep order. They gave, however, a remarkable proof of their steadiness soon afterwards, when the military chest had to be abandoned near Nogales. Casks containing dollars to the value of £25,000 were rolled down the ravine on the right-hand side of the road; the rear regiment of the Reserve were present, but not a man attempted to leave the ranks in search of pickings. They were better off without a load of coin, as some of the camp followers found. The wife of Maloney, the master tailor of the 52nd, burdened herself with such a mass of silver that when the march was over and her turn came to embark, she slipped in stepping on to

1809 the transport and was dragged down and drowned by the weight of her dollars.

At Lugo, Sir John Moore halted and offered battle. But the pursuit was no longer in Napoleon's hands: he had handed the army over to his marshals as soon as he saw that he had lost the race for Benavente. Soult was now in command of an advanced force, and he refused to attack alone. Moore accordingly pushed on for Betanzos; torrents of rain made the march more trying than ever, and the number of stragglers increased so greatly that special measures had to be taken for bringing them on and saving them from the French cavalry. Colonel Cadell, of the 28th, has recorded that near Betanzos on January 9 a sergeant of the 52nd, whose name he does not mention, collected a considerable number of these exhausted men and made a stand against the cavalry, whom he gallantly repulsed. Lieut.-Colonel Hull, commanding the 43rd, has left an account of a precisely similar action performed at the same time and place by Sergeant William Newman of his own regiment. These two accounts may possibly both relate to the same man, or perhaps sergeants from the two regiments stood by one another there as they have done many times before and since. Sergeant Newman of the 43rd was reported to General Fraser, and by him to the Commander-in-Chief, who gave him a commission in the 1st West India Regiment.

The army reached Corunna on January 11, the Reserve staying behind at El Burgo to blow up the bridge. They had to retire hastily on the evening of the 13th, for Soult had at last got his guns up and was shelling them heavily. At Corunna they found the whole army looking out anxiously for the transports, which were not yet in sight. Next morning the French 1809 repaired the bridge at El Burgo and advanced to a strong position for attacking: but the transports entered the harbour the same evening and the embarkation of the artillery, cavalry and baggage, with the sick and wounded, was nearly completed before Soult was ready for battle. At last, on the morning of the 16th, his artillery opened fire, and by two o'clock the light troops on both sides were engaged.

The two armies were very unequally matched: Moore had no cavalry and no artillery but a few six-pounders, while Soult had succeeded in posting a great battery of heavy guns on a hill on his left, and his cavalry, by threatening to outflank the main British position, made it necessary for Moore to keep a division in reserve. The French, moreover, numbered 20,000 men to 14,000. The sole advantage of the English lay in their new arms and ammunition, of which they had found an ample supply in the Spanish stores at Corunna, and in their eager desire to repay their enemy before they left the country to him.

Soult attacked in three columns, and that on the left at once took the village of Elvina on the extreme right front of the British position. Moore, seeing Baird's right in danger of being turned, sent forward from the reserve the 52nd and five companies of the Rifles, who not only drove back the attack but established themselves in the front of the enemy's position. At the same time the 50th and 42nd retook Elvina, and desperate fighting took place all along the line, General Baird himself being severely wounded. Moore too, while watching the result of the struggle for Elvina, was struck on the left breast by a cannon-shot. "The shock," Napier says, "threw him from his horse with

1809 violence: yet he rose again in a sitting posture, his countenance unchanged, and his steadfast eye still fixed upon the regiment engaged in his front, no sigh betraying a sensation of pain. In a few moments, when he saw the troops were gaining ground, his countenance brightened and he suffered himself to be taken to the rear. . . . As the soldiers placed him in a blanket his sword got entangled and the hilt entered the wound: Captain Hardinge, a Staff Officer, attempted to take it off, but the dying man stopped him, saying: 'It is as well as it is: I had rather it should go out of the field with me.'"

His military secretary, John Colborne, who afterwards commanded the 52nd, has given a vivid description of the last scene in a house on the quay, where Moore's friends and his faithful servant François were gathered round him. He was perfectly calm and knew everyone, asking them all, as they came in, for news of the battle and of his own staff. Only when he tried to send a message to his mother he was overcome for a moment and could not go on. Presently he said: "I have made my will and have remembered my servants. Colborne has my will and all my papers." At this moment Colborne came in. "On my entering the room," he says, "the General knew me and spoke most kindly to me and said: 'Colborne, have we beaten the French?' I replied, 'Yes, we have repulsed them at every point.' 'Well,' says he, 'that is a satisfaction. I hope my country will do me justice.'"

Moore then turned to Colonel Anderson and said: "Remember you go to Willoughby Gordon [then Military Secretary to the Duke of York], and tell him that it is my request and that I expect he will give a lieutenant-colonelcy to Major Colborne: he has been

long with me—and I know him to be most worthy of 1809
it." He then said, " Remember me to General Paget—General Edward Paget—he is a fine fellow." Finally, he asked again if all his aides were safe, and begged Stanhope to remember him to his sister, Lady Hester Stanhope, to whom he was deeply attached. Then " he died in a moment after he had spoken, without the least symptom of pain."

His grave was dug that night on a bastion of the citadel, and early next morning Colborne, Anderson, Percy and Stanhope lowered his body into it wrapped only in his martial cloak. The last brigade of his army were embarking in haste : the French guns were booming from the heights above the harbour : the 2nd battalion of the 43rd, who had covered the embarkation, were retiring within the fortress. They held it till night, and then, by the light of two burning ships which had run ashore, they reached the fleet at last and sailed for England.

CHAPTER VII

1809–1811

The Walcheren Expedition—The King's hard bargains—Second period of the Peninsular War—The march to Talavera—Formation of the Light Division—Battles of Almeida and Busaco—Fighting at Redinha, Cazal Novo, and Foz d'Arronce—The 43rd at Sabugal—The battle of Fuentes d'Onor.

1809 ON June 18, 1809, the armament for the ill-fated and ill-famed Walcheren Expedition assembled in the Downs. The naval force consisted of 35 ships of the line, and more than 200 other armed vessels, commanded by Admiral Sir Richard Strachan : the army, under the Earl of Chatham, numbered nearly 40,000 of all ranks, and included a Light Brigade composed of the 2nd battalions of the 43rd, 52nd, and 95th.

There has been nothing in English history like this brief and disastrous campaign : the men died of fever, while their commanders quarrelled.

> "Lord Chatham with his sword undrawn,
> Stood waiting for Sir Richard Strachan;
> Sir Richard, longing to be at 'em,
> Stood waiting for the Earl of Chatham."

Flushing was bombarded on August 13 and surrendered on the 15th : but immediately afterwards the British army went sick. In a fortnight more than 12,000 men were down; 7000 died, and few recovered entirely. The 43rd were re-embarked at two hours' notice on August 30, and when they crawled ashore at Harwich, a rustic was heard to pass the immortal comment, " I say, Bill, there goes the King's hard bargains." 1809

In the meantime, the first battalions of the 43rd, 52nd and 95th had sailed for Portugal on May 29, more than 3000 bayonets strong, under General Robert Craufurd. They landed in the Tagus, and marched to join Sir Arthur Wellesley; slowly at first, then by rapid stages, and lastly by that famous forced march which is one of their great memories. " These troops," says Napier, " after a march of twenty miles were in bivouac on July 28 near Malpartida de Placenzia when the alarm caused by the Spanish fugitives spread to that part. Craufurd, fearing that the army (under Wellesley) was pressed, allowed the men to rest for a few hours, and then, withdrawing about fifty of the weakest from the ranks, commenced his march with the resolution not to halt until he reached the field of battle. As the Brigade advanced, crowds of the runaways were met with. . . . Indignant at this shameful scene, the troops hastened rather than slackened the impetuosity of their pace, and leaving only seventeen stragglers behind, in twenty-six hours crossed the field of battle in a close and compact body: having in that time passed over sixty-two English miles in the hottest season of the year, each man carrying from fifty to sixty pounds' weight upon his shoulders."

The Brigade was loudly cheered by the whole army as it advanced to take up outpost duty immediately. And though, to their great regret, they had come up too late for the battle of Talavera, neither the 43rd nor the 52nd had been unrepresented in the fighting:

1809 for a company of each was on the ground, composed of officers and men who had been invalided at Lisbon from Sir John Moore's force in the previous December. These two companies formed part of General Richard Stewart's brigade on the extreme British left, and distinguished themselves by repulsing a French attack with the bayonet.

The General Order by which the 43rd, 52nd and 95th were to "compose a Light Brigade under command of Brigadier-General Robert Craufurd" bears no actual date, but there can be no doubt that the morrow of Talavera was the birthday of the immortal Light Division. The change of title was made by another General Order of Lord Wellington's, dated February 22, 1810. "The 1st and 2nd[1] battalions of Portuguese Chasseurs (Caçadores) are attached to the brigade of Brigadier-General Craufurd, which is to be called the *Light Division.*" By a further Order of August 4, the Light Division was formed into two brigades, one under Lieut.-Colonel Beckwith, consisting of the 43rd, the 3rd Caçadores, and four companies of the 95th, the other under Lieut.-Colonel Barclay, of the 52nd, the 1st Caçadores, and the remaining four companies of the 95th.

The Division was cantoned on the left bank of the
1810 Coa until March 1810; it was then pushed forward as a corps of observation to watch Marshal Masséna, who was investing Ciudad Rodrigo. The place surrendered on July 10, and the Light Division, whose retirement had already begun, were back on the Coa by the night of the 23rd. Craufurd had positive instructions not to risk an action with the river behind him: but he seems to have thought himself bound to

[1] "2nd" is evidently a clerical error for "3rd."

GENERAL SIR JAMES SHAW KENNEDY, K.C.B.,
ADJUTANT OF THE 43RD, 1809.

hold on as long as possible in order to prevent or delay 1810 the investment of Almeida by Masséna and Ney. The result was that, on the morning of July 24, he found himself in the very dangerous position which Lord Wellington had foreseen. The Light Division was alone with only eight squadrons of cavalry and six guns, between Almeida and the Coa, on a front of a mile and a half; Ney's corps of 20,000 infantry, with 3500 cavalry and 30 guns, was driving in their outposts and manœuvring to cut them off from the river; to make good their retreat they had to cross a succession of ridges and ravines, and then pass in succession over a single bridge in face of an overwhelmingly superior force.

It was evident that the bridge must, if the attack was pushed home, be a deathtrap for one side or the other: and neither French nor English shrank for a moment from the trial. Ney attacked with such violence that it became doubtful whether the British infantry line could be maintained unbroken till the cavalry and artillery had been withdrawn across the bridge. The two brigades, the 43rd on the left and the 52nd on the right, fought every inch of the ground, retiring from rock to rock, followed hotly by the French skirmishers, who had "fine fun," says Lieutenant Henry Booth of the 43rd, "pelting us pretty handsomely down to the river!" An artillery caisson was overturned in the road, but recovered by a charge of a company of the 52nd; another half company was cut off in a stone windmill tower, but lay low until night, and then escaped. The 43rd and 95th suffered severely while covering the bridge, but eventually the Division got across at a cost of 30 killed and 270 wounded and missing. Then came their turn. The

1810 six guns with them were the battery of Horse Artillery under Captain Hew Dalrymple Ross, famous from this day onwards as "the Chestnut Troop"; their commander had the power of making them reach inaccessible positions and fire from impossible platforms. Perched on the highest crags above the Coa, they helped to make the passage of the bridge a truly forlorn hope, and Ney, after losing 600 men in three desperately gallant attempts, withdrew his men at four in the afternoon.

This action was one which ought never to have been fought, and Lord Wellington was seriously displeased with Craufurd. But for the Light Division, which had successfully stood up to a whole French army, he had nothing but praise. In his dispatches he mentioned "Lieut.-Colonels Beckwith, Barclay and Hull (of the 95th, 52nd and 43rd), and all the officers and soldiers of those excellent regiments," and afterwards spoke of this fight as one of the most brilliant of the exploits of the Light Division.

On August 27 Almeida suddenly surrendered, and the Light Division was withdrawn in the direction of Busaco, where Lord Wellington intended to offer battle. The enemy advanced on September 26; that night the 43rd and 52nd, who formed on this occasion the left brigade of the Light Division, were placed in line on a small plateau in front of the convent of Busaco, with rising ground before and behind them. This position, being invisible from below, was virtually an ambush, as the sequel proved.

The battle was a simple one. The Sierra of Busaco is a steep ridge eight miles long: the British and Portuguese held it with about 50,000 men, and Masséna made a most determined frontal attack upon it with

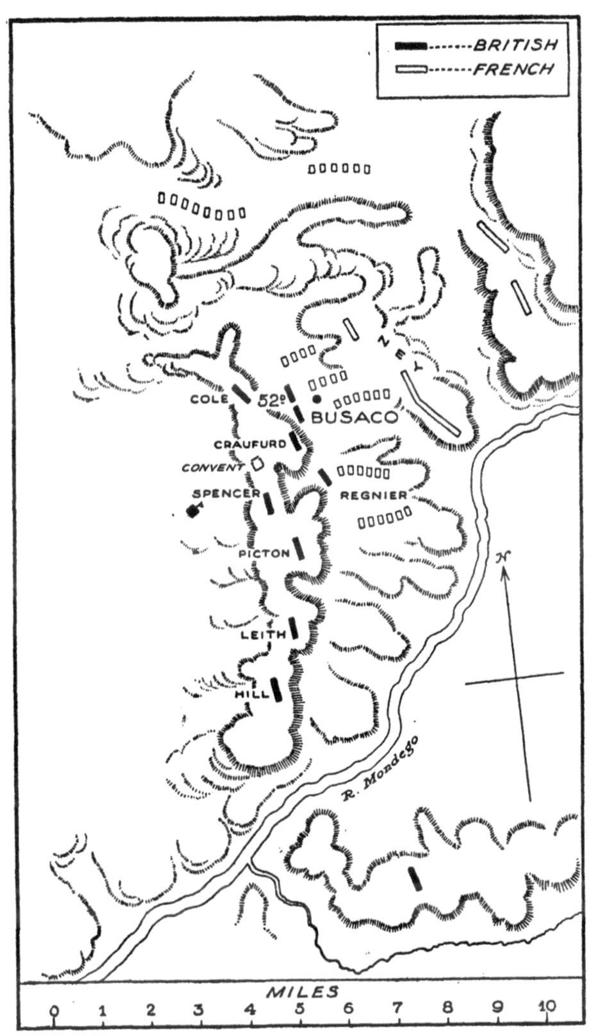

Plan of Battle of Busaco.

1810 65,000. Before dawn on the 27th he planted Ney with three columns opposite the British left, and Regnier with two opposite the right. Regnier had the easier slope to climb, and his attack got in with astonishing rapidity: it broke a Portuguese regiment, pushed back Picton's Division, and established for a moment a hold on the top of the hill. Then two guns enfiladed the columns with grape, and the 88th and 45th regiments charged them so furiously that their right gave way, and French and English rushed together headlong down the slope. The remainder of Regnier's men were then driven off by the 9th and 38th, and when Regnier tried to restore the fight he found it impossible, for Hull's corps had now drawn in, and would have taken him in flank.

Ney in the meantime had also attacked with wonderful impetuosity. He had a harder task than he knew; among the rocks on the edge of the hill were some holes forming natural embrasures, and in these Captain Ross, whose eye no opportunity could escape, had placed his guns to smash the attack with grape and canister. Behind, on the very top of the slope, could be seen some German troops: but the 43rd and 52nd were out of sight in the hollow between. Craufurd himself was standing alone among the rocks, on the look out, and when Ross had withdrawn his guns and the French were shouting victoriously within a few yards of the edge, he gave the signal to the brigade to charge, by waving his hat. Eighteen hundred British bayonets drove straight at the attacking column: its head was thrown violently back upon the rear, both flanks were overlapped, three shattering volleys were poured into them at five yards' distance, and the whole mass went flying down the hill. After this there was a little skirmishing, but the battle was 1810 lost and won : the French had three generals wounded and one killed—General Simon was both wounded and captured by a private of the 52nd. Masséna's total loss was probably 4500, of whom 800 were killed. The British loss was only 1300 in all, and the casualties in the Light Division were marvellously few: the 52nd had three men killed, and three officers and ten men wounded, the 43rd only one officer and eight men wounded. Both regiments were specially mentioned in Lord Wellington's dispatches, and stated to have driven the enemy down " with immense loss."

Wellington now retired to the strongly fortified lines of Torres Vedras; Masséna followed, but by March 5 he found himself obliged to retreat from lack 1811 of supplies, and the Light Division at once moved after him. On March 12 they found the light troops of Ney's rearguard occupying the entrance of the defile and the woods about two miles in front of the village of Redinha, with the main body drawn up on the plain. The Caçadores and the 52nd advanced to clear the wood on the left of the road, " and I must add," said Wellington in his dispatch, " that I have never seen the French infantry driven out from a wood in more gallant style." Captain Mein's company of the 52nd then advanced into the plain, and not only came under the fire of a whole French battalion in line, but was charged by a squadron of dragoons. This and other skilful moves by Ney induced Wellington to hold back the advance until the rear division could be brought up : but when his lines deployed for the attack Ney fired the village and got away to the river under cover of the smoke. He might, perhaps, if harder pressed have been seriously beaten : in

1811 Napier's judgment "Lord Wellington paid him too much respect."

The Light Division was now commanded by Sir William Erskine in the absence of General Craufurd. The 13th was spent in severe skirmishing: on the 14th there was a dense fog, in which the 52nd began the day by passing the French outposts at Casal Novo without either side knowing it, and by almost capturing Ney himself. When the fog rose, the regiment was seen to be right among the enemy, "appearing like a red pimple on the face of the country, black with the French masses!" Captain William Napier with six companies of the 43rd was instantly sent forward to support the 52nd, followed by the whole of the Light Division. They then turned the enemy's right, and when Picton and Cole reached the left flank a confused retreat followed in which two companies of the 52nd, under Captains Douglas and Reynett, enfiladed the fugitives with great effect. The Division lost this day 11 officers and 150 men—among them was Captain William Napier, wounded for the second time in less than eight months, and this time very severely. He had greatly distinguished himself in the action, and was promoted to a brevet-majority.

On the following day, the 15th, Wellington struck one of his quickest and most decisive strokes. The troops reached Ceira late in the day, tired and hungry, for they had outrun their commissariat, and had had no bread for four days. But while they were lighting their bivouac fires their commander had seen his opportunity and ordered an immediate attack of Ney's position, which was near the village of Foz d'Arronce, on the near side of the river. The Light

Division was thrown against the French right, while 1811 the horse artillery opened sharply and suddenly on the left. The attack was instantly successful: Ney's left wing was driven in and fled from their bivouacs to the river in confusion, leaving their provisions and ready boiling kettles to our men. Worse was to come: they were so hotly pressed by Captain Dobbs's and Captain Madden's companies of the 52nd that great numbers were drowned or trampled to death in attempting to cross the bridge: while their own comrades on the other bank mistook them in the darkness and fired into them for a considerable time. For this brilliant little affair Wellington once more thanked both officers and men, and he further requested the commanding officers of the 43rd, 52nd and 95th to name a sergeant of each regiment to receive a commission, in token of his particular approbation of these three regiments. Accordingly Sergeant-Major Kent of the 43rd and Sergeant-Major Mitchell of the 52nd were appointed Ensigns in the 60th and 88th regiments.

On the 25th the Division was reinforced by the arrival from England of the 2nd battalion of the 52nd, who proved themselves at once in the fight with Regnier's corps at Sabugal on the Coa. In this action the Light Division were to cross the river by a ford and turn the French left, while two other divisions attacked their front: but in the fog and rain they failed to find the flank and came up against the enemy's left front before the other divisions had arrived. The 43rd and 95th were at first alone: they charged and routed the columns opposite, but were at once counter-attacked by three fresh columns, fired upon by a howitzer from behind a wall, and threatened by

1811 cavalry. They charged again, took the wall and the howitzer, and held on until the 2nd brigade came up. The 2nd battalion of the 52nd then formed on their right, the remainder formed a second line, and a third charge drove the enemy back once more. Another division then appeared and Regnier retired.

This was one of the 43rd's great days: the victory was mainly due to the skill and coolness of the regiment's fighting, and to the initiative shown by two officers, Lieutenant Hopkins, who with his company seized and held a piece of high ground, preventing the British right from being turned, and Major Patrickson, who ordered the first charge on his own responsibility, while his brigadier was out of touch on the left. The qualities shown were those of the finest Light Infantry. "We have given the French a handsome dressing," wrote Wellington, "and I think they will not say again that we are not a manœuvring army." In his dispatch he said, "I consider the action . . . to be one of the most glorious that British troops were ever engaged in. The 43rd Regiment, under Major Patrickson, particularly distinguished themselves."

Almeida was now invested by the British army, and the French made a strong effort to relieve it. On May 3 they attacked the right of Wellington's position at Fuentes d'Onor, and were only driven off after a sharp struggle. On May 4 Masséna came up himself, and on the 5th he made a second attack in force. One corps he sent against the village of Fuentes, another against Poço Velho, where the Light Division was posted in support of the 7th Division, with all the cavalry available—about 1000 sabres. The 7th Division were soon outflanked and forced out of Poço Velho: Wellington withdrew them behind a ridge on

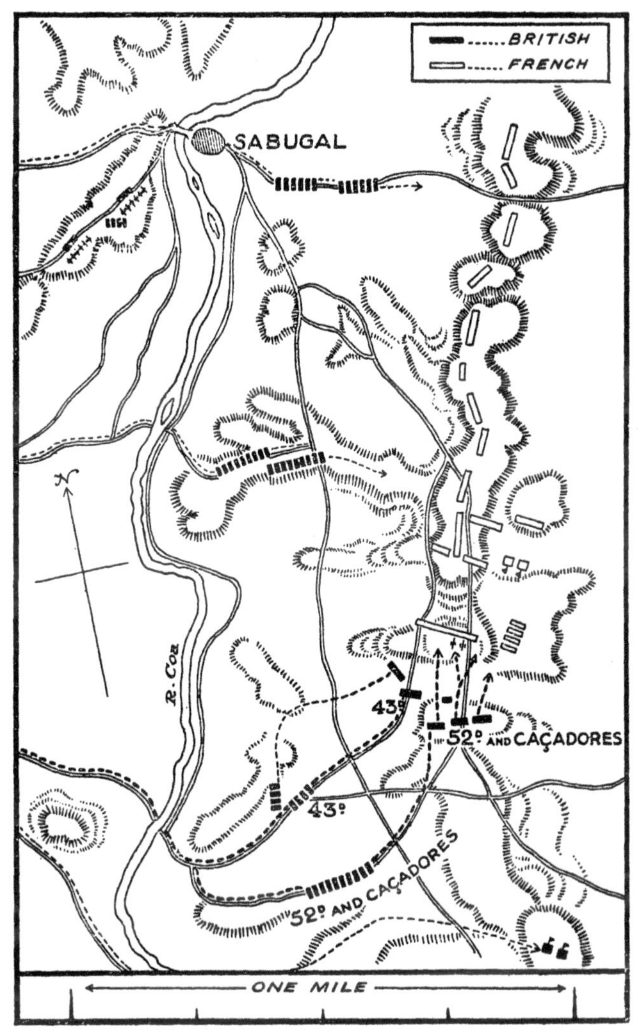

PLAN OF BATTLE OF SABUGAL.

1811 his extreme right, where they could move to a new position, and to prevent their being cut off he ordered the Light Division to retire across the open plain between, formed in squares in echelon of battalions.

Craufurd had just returned to take command, and this was his last action in the field: it was a fitting climax to a thoroughly professional career. "Never perhaps in modern war," says Captain Moorsom, "was a more beautiful movement made, nor at a more critical moment, than by the Light Division on this occasion. The cavalry of Montbrun, numbering 5000 sabres and flushed with their advantage, pressed round the battalion squares without daring to storm them: the French artillery plunged into their close ranks wherever a clear range could be got: and for nearly three miles these veterans held in their conduct the fate of the British army." But in an hour's time they had done their work: the rocks were reached, the guns got into action, and the troops of Masséna broke down before the fire of the Light Division in its new position. Towards evening the 52nd reinforced the village of Fuentes: the whole line was entrenched during the night, and next day Masséna retired, leaving Almeida still invested. He had taken 300 prisoners in his first attack: but his casualties were at least double ours, which amounted to 1500 of all ranks. He was superseded shortly after this action by Marshal Marmont, who was successful in raising the siege of Badajoz. Wellington replied by blockading Ciudad Rodrigo: and during the winter the Light Division were employed in their quarters in making fascines and gabions for the coming siege.

CHAPTER VIII

1812

John Colborne appointed to command the 52nd—He captures San Francisco—The storming of Ciudad Rodrigo—The death of General Craufurd—The capture of Fort Picurina —The 43rd and 52nd at Badajoz—The battle of Salamanca—The retreat from Madrid.

FOR the 52nd the year 1811 had been made memorable 1811
by the appointment to the regiment of the greatest soldier who ever led it into action. Colonel Barclay, who at Busaco had been returned as "slightly wounded," had died within a few months afterwards, and on July 18, 1811, John Colborne was gazetted Lieutenant-Colonel of the 1st battalion. He had already commanded a brigade at Busaco and at Albuera, and he lived to be a Commander-in-Chief, a Field-Marshal and a peer: but the great days of his career were those in which he served with the 52nd.

From August to October Colborne was in England on sick leave: he joined the regiment in November,
and on January 8, 1812, marched with the Light 1812
Division to begin the investment of Ciudad Rodrigo. Before an assault could be made on the fortress itself the outlying redoubt of San Francisco must be captured, and General Craufurd determined to do this by a surprise attack the same night. He appointed Colborne to command, and gave him a detachment consisting of two companies of the 43rd, four of the

1812 52nd, two of the 95th, and others from the Portuguese battalions of the Division.

Colborne detailed his men into three parties and gave his captains most minute and carefully thought-out instructions, warning them especially to move in perfect silence until the moment came for opening fire. The advance guard of four companies was to creep up to the brow of the glacis, from which they could fire straight into the fort across the ditch. Behind them a ladder party was to be ready, and behind these again the men who were actually to scale the walls and lead into the redoubt. As soon as it was dark the advance guard moved silently forward till it was within fifty yards of the fort. Then Colborne, who was with it himself, gave the order "Double quick!" The men rushed for the crest of the glacis, and though the rattling of their canteens as they ran gave the alarm, the defence had only time to fire one round before the advance guard were up the glacis and had opened so hot a fire into the fort that not a man dared show himself. The party with the ladders then came up, got down into the ditch between the glacis and the fort, and placed their ladders against the walls. A few shells and hand grenades came over the parapet, but the storming party rushed up shouting "England and St. George!" and the fort was taken in twenty minutes from the beginning of the attack.

"Thank goodness that's over!" exclaimed Colborne's orderly sergeant MacCurrie. It had been an anxious bit of work for every one. Lord Wellington, Colonel Barnard and General Craufurd were all waiting together in suspense, for in the pitchy darkness they could see nothing of what was happening. Suddenly

JOHN COLBORNE AS A YOUNG MAN.

a great cheer was heard. Colonel Barnard, in his 1812 relief and excitement, flung himself on the ground, whereupon General Craufurd, not seeing who it was, cried out, "What's that drunken man doing?" Soon afterwards a private soldier came up, sent by Colborne to inform Lord Wellington of the success of the enterprise. The man was much excited. "I've taken the fort, sir." Wellington replied, "Oh, you've taken the fort, have you? Well, I'm glad to hear it," and rode off. He was greatly pleased with the night's work, and wrote in his dispatch: "I cannot sufficiently applaud the conduct of Lieut.-Colonel Colborne and of the detachment under his command on this occasion." And Craufurd, who hardly ever applauded any one, was heard to remark, "Colonel Colborne seems to be a steady officer." The complete success of this attack made it appear a simple affair: but it was in truth a striking instance of difficulties overcome by carefully planned and finished work. Its value may be seen from the fact that Colborne achieved in twenty minutes what would by regular siege methods have taken five days: and by daybreak he had consolidated his gain and thrown up a first parallel of 600 yards against the main fortress.

The siege was now pushed forward hurriedly, for Marmont was approaching to relieve the place. The trenches were manned in twenty-four-hour turns by the Light Division and the 1st, 3rd and 4th Divisions. The weather was freezing, and the river Aguada had to be forded going and returning, so that "a pair of iced breeches were usually the accompaniment of each man on twenty-four hours sharp duty." At last on the 19th the assault was ordered, and at nine o'clock that night the Light Division was formed behind

1812 the convent of San Francisco, and shortly afterwards advanced to the attack.

The forlorn hope was led by Lieutenant Gurwood of the 52nd, with 25 volunteers. The storming party followed—100 volunteers from each regiment, those of the 52nd under Captain Joseph Dobbs, those of the 43rd under Captain James Fergusson, and the whole under command of Major George Napier of the 52nd. When these troops entered the ditch they at first mistook the position of the breach, and the forlorn hope diverged to the left and began placing their ladders against the face of a ravelin. But officers of the Royal Engineers had been stationed to guide the attack, and one of them, Lieutenant Elliot, called out, "You are wrong! this is the way to the breach in the *fausse braie!*" Isaac Wild, a private of the 43rd, had by this time actually mounted the false breach, but the whole party sprang down again, and rushed to join the storming party in the great breach. They entered it together, thirty or forty abreast, Lieutenant Steel of the 43rd and Gurwood of the 52nd among the first: a gun was taken, a party of the enemy bayoneted, and the rest driven back into the retrenchment behind.

In the meantime the supporting regiments under Colborne and Craufurd himself had mounted the lesser breach. The 52nd then wheeled to the left and the 43rd to the right, clearing the ramparts, and at the moment when the storming party and the 3rd Division were being held up by the defenders in the retrenchment, the 43rd delivered a determined flank attack which put an end to the resistance. A few minutes afterwards the whole town was in possession of the British. In the citadel Lieutenant Gurwood captured the French governor, General Barrié. Gurwood was wounded, but Lord Wellington summoned him next day to the breach by which he had entered, and there presented him with General Barrié's sword. He also promoted him to a captaincy in the Royal African Corps, from which he exchanged into the 9th Light Dragoons and returned to the Peninsula as Brigade-Major of the Household Cavalry. 1812

The French lost on this occasion 300 killed and 1500 prisoners: the British casualties were 90 officers and 1200 men killed and wounded. Among the dead were Captain Joseph Dobbs of the 52nd and Lieutenant Bramswell of the 43rd: Colonel Colborne and Major George Napier were severely wounded, and General Craufurd, who was shot through the lungs, died five days afterwards. But the Light Division had again done its work: "in the storm," wrote Lord Wellington, "nothing could withstand the gallantry with which these brave officers and troops advanced and accomplished the difficult operation allotted to them, notwithstanding all their leaders had fallen." He particularly mentioned General Craufurd; Colonel Colborne, Major Gibbs and George Napier, and Lieutenant Gurwood, of the 52nd; and Lieut.-Colonel Macleod and Captain Duffy of the 43rd. Majors Gibbs and Napier, with Captain Mein and Lieutenant Woodgate, of the 52nd, were also promoted.

On February 23, 500 men from the 2nd battalion of the 52nd were drafted into the 1st battalion, and on the 25th the skeleton of the 2nd battalion, with some unfit men from the first, marched for Lisbon and embarked for home. On the 27th the 1st battalion moved with the Light Division on Badajoz, to the tune of "St. Patrick's Day in the Morning":

1812 and on the night of March 16 the trenches were opened in a storm of wind and rain. The place was evidently of enormous strength: as a man of the Connaught Rangers said when the 43rd relieved them before daylight on the second day, "Soudadrago was but a flay-bite to this."

On March 19 the enemy made a sortie but were beaten back with loss. On the 25th Fort Picurina, an important outwork, was stormed by the parties going off duty from the trenches, a hundred men of the 52nd under Captain Ewart leading the attack. Captain Ewart and his subaltern, Ensign Nixon, were wounded and thirty-four of their hundred men were wounded or killed in this affair, but they had gained an important position for the breaching batteries, and within twelve days afterwards three breaches were reported practicable.

The final assault was ordered for the night of April 6. At 8.30 p.m. the Light Division was formed, and the roll called in an undertone. The attack was to be made on three main points: General Picton with the 3rd Division was to escalade the castle, a detachment of the Guards under Major Wilson of the 48th was to attack the ravelin of San Roque, while the 4th Division, under General Colville, and the Light Division, under Colonel Barnard, were to storm the breaches in the bastions of La Trinidad and Santa Maria. The 5th Division relieved the 4th in the trenches and made feints on some of the outworks.

A little before ten o'clock the move began in complete darkness and silence. The Light Division advanced in columns of sections: first the forlorn hope with ladders and Engineer officers, then the grassbag, axe, and crowbar men, then the storming parties of 100 from each regiment, then the regiments themselves. As the town clock struck ten and the sentries cried their "Sentinelle, garde à vous" a fireball rose into the air and fell near the crowbar men, who at once extinguished it, and the whole force crept forward again. The 43rd, 52nd and part of the 95th gradually closed to columns of quarter distance: they placed their ladders on the edge of the ditch, and began to descend, Captain Fergusson of the 43rd and Ensign Gawler of the 52nd leading their men, with Engineer-Lieutenant de Salaberry. Gawler with some dozen men was already in the ditch, when suddenly there was a tremendous explosion at the foot of the trenches, a blinding blaze of light which exposed the whole attack, and a murderous fire from the head of the breach which levelled everything before it. Fergusson was wounded in the head, and his party annihilated. 1812

But the Light Division was not checked for a moment. Some rushed down the ladders, others flung themselves down, only to lie broken below or to drown in the inundations; others scrambled down by narrow ramps: at the bottom they gathered in masses upon the glacis, a desperately gallant but no longer disciplined mob, making charge after charge at impregnable defences. The breaches were ablaze with fireballs and tar barrels, bursting with shells and powder-casks, barred with planks chained to the ground and studded with large nails. At the top was a close row of *chevaux de frise* set with sharp sword blades, and behind these were picked troops to whom freshly-loaded muskets were incessantly passed up from behind.

The position was desperate: Captain Currie of

1812 the 52nd climbed out and found Lord Wellington, who asked him anxiously, "Can they not get in?" Currie begged for reinforcements: he was given a battalion, but they too were swamped in the confusion. The reserve bugles were sent to the crest of the glacis to sound the retreat: the desperate men in the ditch refused to believe the order genuine. Lieutenant Shaw of the 43rd collected 70 men and attempted a last rush at the left-hand breach. His whole party was laid low by two rounds of grape, and he alone remained standing in the dense smoke. Sullenly the Light Division obeyed the order to withdraw: as the last of them recrossed the glacis the town clock was heard striking twelve.

By this time Picton had taken the castle, and Leith the bastion of San Vincente: the 3rd Division entered the town, and the sack of Badajoz followed: a riot of horror against which the British officers fought in vain. The men were maddened by their losses and sufferings: but they were none the less punished with the gallows, and condemned by all their countrymen since. Seven generals, 300 other officers, and 5000 men fell that night in or around the breaches: the Light Division lost nearly half their entire force. The 43rd and 52nd suffered most heavily of all: the losses of the 43rd were 20 officers and 335 men, those of the 52nd were almost exactly the same—20 officers and 334 men. Among the dead was Lieut.-Colonel Macleod, the brilliant young commander of the 43rd, of whom Lord Wellington wrote that he was "an ornament to his profession, and was capable of rendering the most important services to his country."

In the meantime Marmont had entered Portugal,

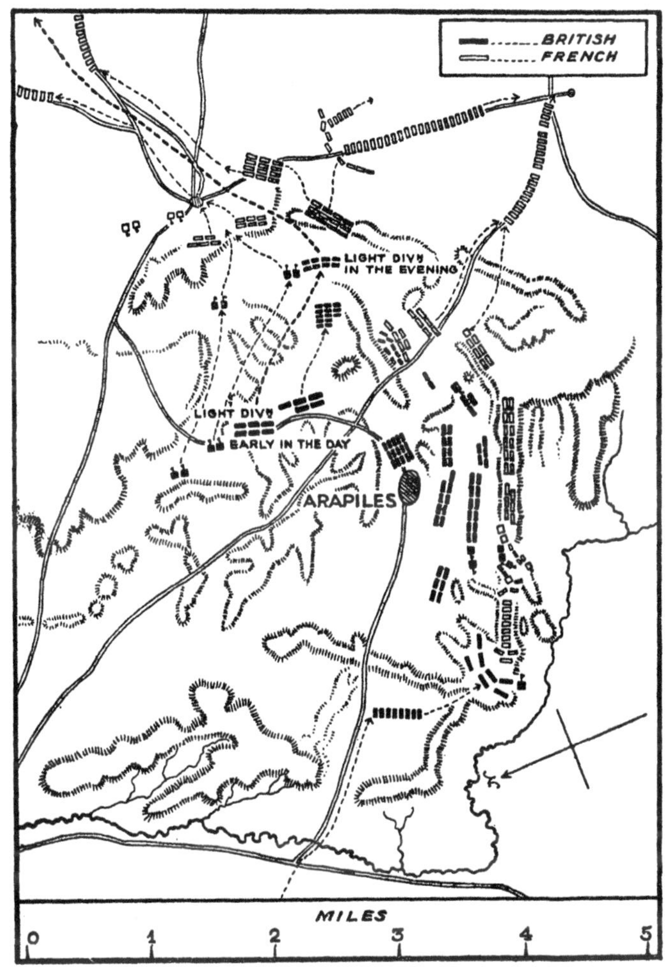

PLAN OF BATTLE OF SALAMANCA.

1812 and after a month's manœuvring he made a sudden attempt on July 22 to cut the British line of communication, moving on Salamanca from the south and east. Wellington attacked at once—his first attack in a general action. On the south his right, under Pakenham, was quickly victorious: in the centre Cole, Leith and Pack were less immediately successful. At 6 p.m. the battle was at its height and dusk was falling, when the Light Division, who had been in reserve on the left, were ordered to attack Foy's division and seize the ford of Huerta to the east. The 43rd led the column and "made a very extraordinary advance in line for a distance of three miles, under a cannonade which though not heavy was constant, with as clear and firm a line as at a review." This was the more notable because it was now dark, and the regiment kept its line simply by touch to the centre: Captain Shaw (Kennedy) who commanded the left-centre company, declared that the line was so well kept as to have been able at any moment to fire a volley and charge with the bayonet.

The enemy retired with a loss of 12,000 men, and were pursued further on the 23rd by the 52nd, followed by the whole Light Division. On August 6 the army marched upon Madrid, which was occupied on the 12th, and held till the end of October. Lack of money and supplies then compelled Lord Wellington to fall back before Soult's advance. On November 12 he offered battle in his old position at Salamanca, but Soult was too wary, and the retreat was continued, covered by the Light Division as rearguard. On the 15th the French cavalry stole through a wood and made a successful raid on some baggage; they also captured General the Hon. Edward Paget, and drove in two of Captain Ross's guns. Lord Wellington 1812 himself galloped up, surveyed the position, and carried the Light Division forward across the ford of the Huerta, riding up the bank in front of the 1st company of the 43rd. The 52nd drove off a scarlet Swiss regiment, who were at first mistaken for friends, and by three o'clock in the afternoon the army was clear. The retreat continued until Ciudad Rodrigo was reached on the 19th, and was declared by some veterans to have been, though shorter than the retreat to Corunna, quite as dangerous and exhausting. The army then went into cantonments for the winter.

CHAPTER IX

1813

The battle of Vittoria—Soult's return—The Pyrenees—Wellington's letter to the Light Division—The storming of San Sebastian—The heights of Vera—Colborne's bluff—The invasion of France—The battle of Nivelle—Colborne bluffs again—The passage of the Nive—Soult's last attack.

1813 THE campaign of 1812 ended with a retreat, and caused much disappointment at home, but it had prepared the way for success, and when Napoleon, after his Russian disaster, recalled Soult with 20,000 men, Wellington's chance came at last. He used it in so masterly a fashion that in the first five weeks of the campaign of 1813 he drove the main French army from the Portuguese frontier to the Pyrenees, and by the middle of October he had established himself in the plain of France.

Operations began on May 20, when the Light Division, which had been reviewed by Wellington on the 16th, marched with the right wing of the army towards Burgos. Sir Thomas Graham joined with the right wing on June 2; on the 13th the French blew up and abandoned the castle of Burgos, and on the 18th the Light Division overtook two of their brigades, one of which was immediately charged and broken up by the 52nd, supported by the 2nd battalion of the 95th. The 43rd defeated the other and captured the valuable stores of its medical staff.

The French, under Marshal Jourdain, but nominally

commanded by King Joseph Buonaparte, were found 1813 on the 20th in position in front of the town of Vittoria, and on the 21st Wellington attacked them there. The position was a strong one, with the long ridge of the Pueblas mountain on the left, and the river Zadorra both in front and all along the right flank: but it had one dangerous defect—in case of defeat the roads across and along the river to the north would be certainly blocked, and the only line of escape would be the easterly route by Salvatierra, which was quite insufficient for an army of any size. This danger Wellington converted into an overwhelming disaster. He sent Sir Thomas Graham's corps to turn the enemy's right and rear, and so close the high road, while he attacked their front. He placed the Light Division in the centre of his line, the 3rd and 7th on the left, and the 4th and 2nd on the right.

The British left and centre had to begin by passing the river. Wellington himself led the Light Division, by a concealed road, to some wooded ground opposite the deep pocket where the river begins its bend from west to south and south-west. Kempt's Brigade crossed first, on the left of the pocket, and the 43rd found themselves actually in the advanced centre of the French position, with King Joseph himself and 5000 troops close upon them. Fortunately the enemy hesitated, Sir James Kempt called up the 15th Hussars to secure his flanks, and in half-an-hour the advance of the 3rd Division began to take effect. Their movement was supported by Vandeleur's Brigade of the Light Division, which rushed a bridge on the right of the river bend, crossed Kempt's track at right angles, and carried the heights of Marguerita on the French right, the 52nd leading the charge.

1813 Before this combined attack the French fell back to a second position, and the 3rd Division deployed into line and attacked again, supported this time by the 43rd. Graham was now in action, but not yet able to force the bridges at Ariaga and Gamarra, and Vandeleur's Brigade therefore moved to his aid along the near side of the river. The 52nd, here finding

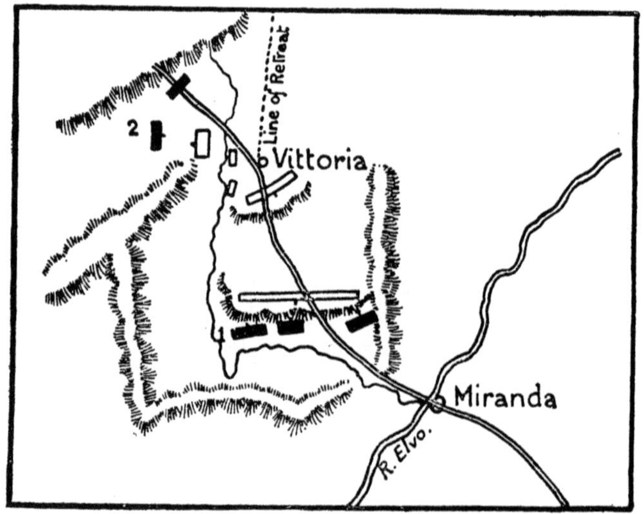

BATTLE OF VITTORIA. JUNE 21, 1813.
1, *Main attack*; 2, *Graham closing the high road. French divisions in outline: British in black.*

themselves exposed to the fire of six guns on a hill in front of them, formed line with the accuracy of a parade, charged the hill and took the battery. "A beautiful line was formed," says Captain John Dobbs, "the enemy's balls knocking a file out of it at every discharge, the sergeants in rear calling out, 'Who got that?' and entering the names on their list of

casualties." Outflanked by this advance, the French 1813
retired to a third position within a mile of Vittoria, where they put a good face on it, and for a short time seemed to be holding their own.

It was now past six o'clock, and Lord Wellington, determined to finish the battle before dark, ordered the 4th Division to attack on the right of the centre. With an irresistible rush they broke the dense column opposed to them, and the whole French line immediately broke up and streamed away in uncontrollable flight. The high road being by this time completely closed, the Salvatierra road was the only way of escape, and it was instantly choked with a fantastic crowd of wagons, carriages, wounded men, cattle and sheep, women and children, ammunition and stores of every kind. The booty included King Joseph's carriages, the military chest with nearly a million of gold, and all the papers and equipment of the General Staff, who were reduced to complete destitution. The pursuit of the troops was kept up by Vandeleur's Brigade, the 52nd finally halting for the night more than three miles beyond Vittoria.

The British casualties were 3300, the Spanish and Portuguese 1600, the French about 6000. But these figures do not tell the tale: Wellington had taken the whole of the enemy's artillery and baggage, and driven them from the field, a rout only paralleled by the *sauve-qui-peut* of Waterloo.

The blow was struck just in time. Soult had been sent back to Spain by Napoleon directly he heard of Wellington's movement across the Ebro; he reached his headquarters on July 13, and at once collected a force of 60,000 men for the relief of Pamplona and San Sebastian. On July 25 he attempted the passes

1813 of the Pyrenees, but was beaten back, with a total loss of 15,000 men, in a series of actions in which Wellington outmanœuvred him and the British divisions outfought his men. By August 2 he was taking breath behind the line of the Bidassoa. On the 30th he sent a force across the river from the heights above Vera with the object of breaking through to San Sebastian and bringing away the garrison: but the Spaniards stood fast, and the French columns in returning found the river swollen and the ford impassable. At the Lesaca bridge they got across in the dark, but suffered heavily from the fire of the 95th and 52nd. The British assault on San Sebastian was ordered for noon on the following day.

As one attack had already failed, Lord Wellington was anxious about this, and decided to add to the 5th Division a storming party of 750 volunteers from the 1st and Light Divisions. His letter to Baron Alten, now commanding the Light Division, was complimentary but perhaps indiscreet: in it he begged him to "send 1 field officer, 2 captains, 4 subalterns, and 100 men, to show the 5th Division how to mount a breach." In the Light Division this caused great enthusiasm, and whole companies volunteered: in the 5th there was hot indignation, and the men threatened to bayonet the volunteers if they found them in front. Their commander, General Leith, solved the difficulty by accepting the volunteers and distributing them along the trenches of the hornwork.

At 11 a.m. the stormers of the 5th Division made their attack and failed. Their 2nd brigade was then sent in with the volunteers from the Light Division: the batteries had been firing for half-an-hour over their heads, and the defence had weakened. By 1813 desperate fighting the volunteers secured a foothold: an explosion then killed many of the defenders, and the Portuguese 13th regiment, under Major Snodgrass of the 52nd, carried the lesser breach, while the British forced their way in on the left. The town was taken by three o'clock: the castle surrendered eight days later.

Soult now built redoubts on the heights above Vera, and made the position very formidable. On October 7 Wellington crossed the Bidassoa to dislodge him: the attack on the forts was entrusted almost entirely to the Light Division, the 43rd and 52nd delivering repeated assaults in column, while the 95th and Caçadores acted as skirmishers. Colborne commanded the 2nd Brigade in General Skerrett's absence, and wrote the following account of the affair. "The Rifles being the first to attack the fort, the French mistook them for Portuguese Caçadores, and rushing out of the redoubt drove them back, so that they all came tumbling on the 52nd. The French were excessively astonished when they saw the red coats behind the Rifles. The Adjutant of the 52nd was surprised to find we were so near the fort. 'Why, sir, we are close to the fort!' 'To be sure we are,' I said, 'and now we must charge.' I then led the 52nd on to a most successful charge, to the admiration of Lord Wellington and others who were watching from another hill. At this moment Sir James Kempt, who was leading the 1st Brigade of the Light Division to a simultaneous attack on the right of the town of Vera, a mile or two off, sent to General Alten to know if the 52nd could not render him some assistance. 'Colonel Colborne give him some assistance!' said

1813 Alten; 'if he could see the hill Colborne's brigade is on he'd see that Colborne has quite enough to do himself.'"

The French, thrown into confusion by this charge, retreated to the next fort, and were again charged and driven out. "After this," says Colborne, "leaving my column I rode on alone with the present Sir Harry Smith into France." On the way, Sir Harry Smith records, he found a force of 400 French troops at the bottom of a ravine, and bluffed them into surrendering by giving orders to imaginary British forces over the top, where a few of Kempt's riflemen were visible. "By this time our men had got well out of the Pyrenees into the plains of France below . . . The prisoners were sent to the rear under charge of a Lieutenant Cargill, of the 52nd, a manly, rough young subaltern, who, on his march just at dark, met the Duke,[1] who says, 'Hulloa, sir, where did you get those fellows?' 'In France—Colonel Colborne's Brigade took them.' 'How the devil do you know it was France?' 'Because I saw a lot of our fellows coming into the column just before I left, with pigs and poultry which we had not on the Spanish side.' The Duke turned hastily away without saying a word. The next morning Mr. Cargill reported this to Colonel Colborne, whom I hardly ever saw so angry. 'Why, Mr. Cargill, you were not such a blockhead as to tell the Duke *that*, were you?' In very broad Scotch, 'What for no? It was fact as death.' It did not escape the Duke, who spoke to Colborne, saying, 'Though your Brigade have even more than usually distinguished themselves, we must respect the property of the country.'"

[1] Wellington was "the Duke" when this was written, but not at the date of the events related.

Colborne tried to excuse his men. "'Ah, ah!' says my lord, 'stop it in future, Colborne.'" 1813

The 43rd on this day did their work equally well, taking two hills in succession and chasing the enemy for nearly five miles down into France. Their losses were not so heavy as those of the 52nd, for they had the shelter of ravines; but the climbing was long and exhausting, and for the first time in the war they left their packs behind for the attack, and only recovered them during the night when fatigue parties came up from Santa Barbara.

The Division distinguished itself on two more occasions before the year closed. At the battle of Nivelle, on November 9, it was placed in the centre of the line, and when the sun cleared the mountain tops was already within 300 yards of the advanced French position on the height of La Petite Rhune. The Rifles moved off to the left; General Kempt mounted his horse, saying, "Now 43rd, let me see what you will do this morning." The regiment took one position after another, and ended by storming first a stone redoubt and then a line of trenches.

The 52nd, by creeping under a wood, cleverly outflanked the enemy's right, passed behind the now abandoned forts of La Petite Rhune, and came up against a strong star-shaped redoubt garrisoned by the French 88th Regiment. The attack up a long exposed slope cost them 200 men and appeared to have failed—the regiment was driven to take cover in a ravine, and could not be got further. General Alten sent Harry Smith to call them off; but Smith's horse was shot, and he failed to reach them. Colborne saved the situation by sounding the bugle for a parley, advancing alone to the redoubt and summoning it to

1813 surrender. The French commander was indignant, but afraid of falling into the hands of the Spanish troops whom he saw in the distance, and ended by marching out with the honours of war.

The campaign closed with three days' brilliant fighting on the Nive, when Soult, after Wellington had forced the passage of the river, attacked him for the last time. He was decisively repulsed on December 10, 11 and 12, and the 43rd and 52nd gained one more honour for their regimental colours. On the 13th Soult retired towards Bayonne, and on the 14th the Light Division went into winter quarters.

CHAPTER X

1814

The 2nd battalion of the 52nd in Holland—The bombardment of Antwerp—The 1st battalion at Orthez—Retrieves a lost battle—Toulouse, the last fight of the war—Wellington's trump card, the Light Division.

At the beginning of the year 1814 the 52nd once more 1814
had two battalions in the field, but this time in two different campaigns. On December 9, 1813, while the 1st battalion was marching out to engage Soult on the Nive, the 2nd battalion was embarking at Ramsgate for active service in Holland. The army there was under the command of Sir Thomas Graham, and the arrival of the 52nd brought him some regimental officers who had shared with him many hard and glorious days in the Peninsula. The battalion was commanded by Lieut.-Colonel Edward Gibbs, who had distinguished himself at Ciudad Rodrigo, at Badajoz, where he lost an eye, and at Vittoria; and among the captains were Charles Diggle and Frederick Love, whose names are still freshly remembered, the one for coolness, the other for unfailing vivacity. They were fortunate, too, in being attached to a Light Brigade which included a detachment of the 95th; and they had for brigadier General Kenneth Mackenzie, who, as Lieut.-Colonel of the regiment, had first trained the 52nd at Shorncliffe, under the eye of Sir John Moore.

1814 The campaign was a short one. On January 13 the French were driven into Antwerp with considerable loss, and the town was bombarded both then and on several days in February with the object of destroying the enemy's fleet, which was frozen up in the basins. The 52nd were commanded in the second attack by Captain Diggle, who records that he saw "his late Majesty, King William the Fourth, then Duke of Clarence, riding about the village of Merxem, the skirts of his great coat perforated by a bullet, and wholly regardless of danger, as is the wont of the Royal Family." On another occasion the Duke of Clarence had a still narrower escape. He was in a new battery talking to an officer of the 52nd, whose men formed the battery guard; the enemy became aware of his presence and prepared to fire, but the officer of the 52nd, seeing them removing their mantlets, persuaded his Royal Highness to retire and ordered the detachment to take cover outside. Immediately afterwards the battery was completely ruined by the enemy's salvoes.

The guard of the 52nd escaped without loss, "owing to this foresight of a cool Peninsular officer," and their good fortune held on every occasion on which they were under fire. Captain Diggle's horse was struck down, as he stood beside it, by a cannon ball; Captain Love was knocked over by the frozen earth thrown up from a shell crater; guns were dismantled near them and artillerymen killed, but in three months the regiment had not a single casualty to record. The campaign ended abruptly with the abdication of Napoleon; and for the remainder of the year the battalion was stationed successively at Brussels, Antwerp, Tournai and Ypres.

In France, too, the war was now drawing rapidly

to an end. The Light Division left its cantonments on February 16; on the 25th it reached the heights near Orthez, and halted near the bridge over the river Gare de Pau, opposite the town, which formed the left of the French position. After an attempt to cross higher up, the Division was counter-marched down stream and ordered to strengthen the British left, which was attacking the enemy on the fortified ridges of St. Boes. The battle was again to be a series of climbs and charges as at Vera, but this time the Light Division was like a man going into a fight with his right hand injured, for the 43rd had been sent back to Ustaritz to get their new clothing, which had just arrived from England. 1814

The day began badly. On the British left, Cole's leading regiments, after gaining a foothold in St. Boes, were driven out again and enfiladed by artillery; in the centre the 3rd Division were also for a time repulsed. The Light Division was between these two attacks, on a spur running up to join the main ridge at the exact point where Foy was successfully pushing back Cole. The Caçadores were thrown forward, but they too failed; General Alten came riding over to the 52nd, and said, "Now, Colborne, you go on and attack."

Wellington had now come up, and was standing dismounted on a knoll. As Colborne passed him he called out, "Hello, Colborne, ride on and see if artillery can pass there." A piece of marshy ground lay on the right of the road. Colborne galloped into it, and came back to say that it was passable. "Well, then, make haste; take your regiment on and deploy into the plain. I leave it to your disposition." Before leaving the spur Colborne met Sir Lowry Cole, much excited. "Well, Colborne, what's to be done? Here we are,

1814 all coming back again as fast as we can." "Have patience," said Colborne, "and we shall see what's to be done." He then ordered the regiment to cut across the marsh and go straight up again at the main ridge. "They did it beautifully," he said afterwards. "When all the rest were in confusion, the 52nd marched down as evenly and regularly as if on parade. The French were keeping up a heavy fire, but fortunately the balls all passed over our heads. I rode to the top of the hill and waved my cap, and though the men were over their knees in the marsh, they trotted up in the finest order. As soon as they got to the top of the hill I ordered them to halt and open fire. . . . We were soon supported by the other divisions, and the French were dispersed."

Wellington rode up behind and saw it all. "This attack," he said in his dispatch, "led by the 52nd Regiment, dislodged the enemy from the heights and gave us the victory." It was a bad disappointment for Soult: the story goes that at the dangerous moment he had slapped his thigh, and exclaimed, "At last I have him!" Cole and Picton had failed, but till the Light Division had been thrown in it was too soon to hope, and afterwards it was—then and always—too late.

In the opinion of Sir James Shaw Kennedy of the 43rd, this attack was one of "the two most brilliant events of Lord Seaton's life." Sir Harry Smith says that Colborne was at this time, in the judgment of the whole army, inferior to none but Wellington himself. Certainly he was the most perfect instrument his Commander-in-Chief ever wielded, for the instinct of the two always followed the same line. They had, too, a certain sensible plainness of manner in common. At

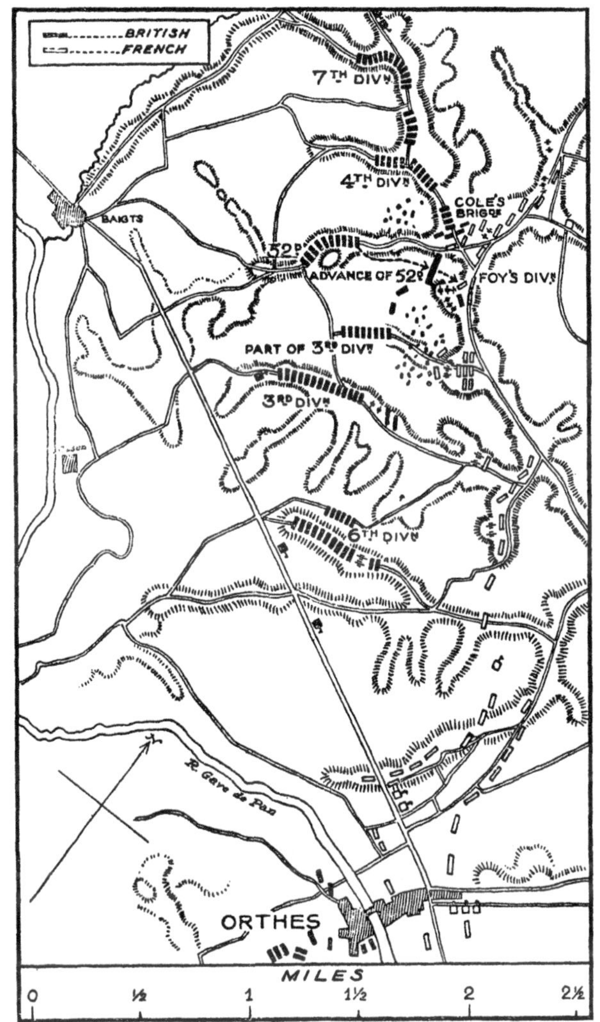

PLAN OF BATTLE OF ORTHEZ.

1814 the climax of the battle of Orthez, Major George Napier came up to his colonel with a face of grave concern, saying, " Poor March [1] is wounded." " Well," replied Colborne, " I can't help it. Have him carried off." Napier must almost have thought he heard Wellington speaking.

On the night of the 28th the 43rd regained touch with the army near Orthez, and learned to their disappointment that they were too late for the battle. On the following day they marched early, hoping to get ahead of the 2nd Division, but they were blocked intentionally until Wellington happened to come up. " Forty-third," he called out, " what are you doing here? " and in ten minutes he had halted the whole army to let them take their proper place. At St. Sever they were sent forward with Ross's brigade of Horse Artillery to take possession of Mont de Marsan. The inhabitants flocked out to see them enter, and were astonished to see them so poorly clad, with worn tunics and trousers made out of blankets. They had, as we have seen, just received new uniforms, but these they were economically carrying on the top of their packs!

The battle of Toulouse, the last of the war, was fought on April 10, and after peace had been signed, though neither army was aware of the fact. Colborne described it as " the worst arranged battle that could be, nothing but mistakes. . . . It was a most extraordinary battle. I think the Duke almost deserved to have been beaten." The Light Division was to support the main attack by the Spanish troops on the heights of La Pugade. The position was very strong, the Spaniards were repulsed, and the Light Division 2nd

[1] The Earl of March, afterwards Duke of Richmond, was in command of the leading company of the 52nd in the attack.

THE DUKE OF RICHMOND, K.G.,
CAPTAIN AND BREVET MAJOR IN THE 52ND, 1815.

Brigade checked the French pursuit. In the afternoon 1814
they supported a second attack by Cole's and Clinton's divisions, which was successful, but not until Marshal Beresford had made a costly movement across the enemy's front to take the place vacated by the Spaniards.

"When the battle began," says Colborne, "the Spaniards were sent up a hill to attack the French, who were at the top. It was a most difficult thing. I should have been sorry to do it with two Light Divisions, and I remember standing at the bottom, looking at them with wonder and trembling, and then seeing them come running down as hard as they could . . . they ought never to have been set to do such a difficult thing. . . . When the Spaniards came back, Lord Wellington said to Pakenham: 'There I am, with nothing between me and the enemy!' Pakenham said, 'Well, I suppose you'll order up the Light Division now?' And he replied, 'I'll be *hanged* if I do!'"

Many superlatives and much real eloquence have been spent upon the work of the Light Division, but no evidence of its value could be more convincing than this conversation. To lead an assault, to cover a dangerous movement, or to fight a rearguard action—these were the special duties of a corps of Light Infantry, and they were carried out with perfect and unvarying skill. But in the 43rd and 52nd these two brigades also possessed regiments capable of being used as "firm battalions," and actually so used and relied upon in every moment of difficulty. They were Wellington's best for offence or defence, his first thought and his last resource; they were so instinctively his strength that in the eyes of his staff they had almost become his weakness.

CHAPTER XI

1815

The American War—The 43rd reinforce Pakenham—The attack on New Orleans—Lambert's brigade alone undefeated—Return to England—Napoleon's escape from Elba—The 43rd ordered to Flanders—They arrive the day after the battle—The Duke and the phantom regiment at Waterloo.

1815 AFTER three months' rest at home the 1st battalion of the 43rd was ordered out again to the American War. A thousand and fifty bayonets strong, they embarked under command of Colonel Patrickson on October 10, and on New Year's Day 1815 their transports joined the British fleet in Lac Borgne. The war had already lasted two and a half years with varying fortune: Canada had been twice invaded and twice recovered; the Americans had been defeated at Bladensburg and successful on the Lakes: and the British Government had now decided to attack New Orleans.

When the 43rd joined the expedition the position of the army there was not so good as it had been. The first landing under Sir John Keane on December 22 had taken the enemy by surprise, but the opportunity was not grasped, and Jackson, the American general, took full advantage of this mistake. He brought two 14-gun sloops down the Mississippi which sent round shot and grape among the bivouac fires of the British troops, scattering arms, kettles and burning logs, and causing many casualties. A subsequent 1815 attack was beaten off after severe hand to hand fighting, but again Keane failed to follow up his success, and Jackson then set to work to barricade his position on the high road in front of New Orleans by digging a wet ditch three-quarters of a mile long and building a breastwork of sugar casks and cotton bales behind it.

When the Commander-in-Chief, Sir Edward Pakenham, arrived on Christmas Day he saw that the situation was serious. A reconnaissance on the 28th cost him fifty men and confirmed his opinion. He was compelled to wait for reinforcements: in the meantime the gunners succeeded in destroying one of the enemy's two sloops, and in constructing several batteries behind breastworks of sugar barrels—the soil was too water-logged for digging. At daybreak on January 1, 1815, the army was drawn up for attack and the guns opened fire. The enemy's batteries at once replied with crushing effect, and Sir Edward, to his great mortification, found himself obliged to withdraw his troops. The guns were left with pickets to guard them, and two very unwilling regiments were sent out during the night to bring them back.

On January 5, Lambert's brigade arrived—the 43rd and the 7th Fusiliers. Sir Harry Smith, then Assistant Adjutant-General to Pakenham, describes the relief of the army, "Two such corps would turn the tide of a general action. We were rejoiced!" A grand attack was ordered for the 8th: preparations were made, by damming a little side stream, for throwing the 85th, under Colonel Thornton, across the river, and on the night of the 7th two hundred men of the 43rd were detailed to repair and strengthen a battery

1815 on the right under cover of darkness. Another company of the 43rd, with two more of the 7th and 9th, were told off to storm the Crescent Battery, in which the Americans had now mounted twenty guns. The bulk of Lambert's brigade, " the *élite* 7th Fusiliers and 43rd," were held in reserve : the assault was to be made by 6000 men in three columns. Harry Smith's account of the action is full of vivid detail and also of instructive lessons in the art of war. " About half-an-hour before daylight, while I was with General Lambert's column, standing ready, Sir Edward Pakenham sent for me. I was soon with him. He was greatly agitated. ' Smith, most Commanders-in-Chief have many difficulties to contend with, but surely none like mine. The dam, as you heard me say it would, gave way, and Thornton's people will be of no use whatever in the general attack.' I said : ' So impressed have you ever been, so obvious is it in every military point of view, we should possess the right bank of the river, and thus enfilade and divert the attention of the enemy—there is still time before daylight to retire the columns now. We are under the enemy's fire as soon as discovered.' He says : ' This may be, but I have twice deferred the attack. We are strong in numbers now, comparatively. It will cost more men and the assault must be made.' I again urged delay. While we were talking the streaks of daylight began to appear, although the morning was dull, close and heavy, the clouds almost touching the ground.

" He said : ' Smith, order the rocket to be fired.' I again ventured to plead the cause of delay. He said, and very justly, ' It is now too late : the columns would be visible to the enemy before they could move out of fire, and would lose more men than it is to be 1815 hoped they will in the attack. Fire the rocket, I say, and go to Lambert.' This was done. I had reached Lambert just as the stillness of death and anticipation (for I really believe the enemy was aware of our proximity to their position) was broken by the firing of the rocket. The rocket was hardly in the air before a rush of our troops was met by the most murderous and destructive fire of all arms ever poured upon column.

"Sir Edward Pakenham galloped past me with all his Staff, saying: 'That's a terrific fire, Lambert.' I knew nothing of my general then except that he was a most gentlemanlike, amiable fellow, and I had seen him lead his brigade at Toulouse in the order of a review of his Household Troops in Hyde Park. I said: 'In twenty-five minutes, General, you will command the army. Sir Edward Pakenham will be wounded and incapable or killed. The troops do not get on a step. He will be at the head of the first brigade he comes to and what I say will occur.' A few seconds verified my words. Tylden came wildly up to tell the melancholy truth, saying: 'Sir Edward Pakenham is killed. You command the army and your brigade must move on immediately.' I said: 'If Sir Edward Pakenham is killed, Sir John Lambert commands, and will judge of what is to be done.' I saw the attack had irretrievably failed. The troops were beat back and going at a tolerable pace too; so much so I thought the enemy had made a sortie in pursuit, as so overpowering a superiority of numbers would have induced the French to do. 'May I order your brigade, sir, to form line to cover a most irregular retreat, to apply no other term to it, until you see what has actually

1815 occurred to the attacking columns?' He assented and sent me and other Staff Officers in different directions to ascertain our condition. It was—summed up in a few words—that every attack had failed: the Commander-in-Chief and General Gibbs and Colonel Renny killed; General Keane most severely wounded; and the columns literally destroyed. The column for the right bank were seen to be still in their boats, and not the slightest impression had been made on the enemy."

In this appalling scene of disaster there was only one element of hope left. Lambert's brigade stood unbroken, the Fusiliers within 600 yards of the enemy's lines, the 43rd closely echeloned upon them. "Look," cried Lieutenant Rowley Hill, "look at the 7th and the 43rd, like two seventy-fours becalmed. Why were they not led on?" Lambert himself had been moved by the same impulse, but Harry Smith had coolly reminded him that if he lost he lost his army; if he succeeded he would have no troops left to occupy New Orleans. He withdrew accordingly, and he was probably justified as Commander-in-Chief. But so, too, was Rowley Hill justified as an officer of the 43rd, for he knew what his regiment could do, and what two companies of it had already done that day. The two hundred men of the party working on the battery had run the gauntlet from the left to the centre at the hottest moment of the cross-fire in their eagerness to be with their brigade for the attack. The company told off to storm the Crescent Battery had rushed in with the parties of the 7th and 93rd, and in spite of a murderous fire had forced their way into the ditch and taken four guns. The detachment had lost 8 officers and 180 men out of a force of 240,

but they held on till the whole attack had failed. 1815
The 85th, too, had good reason to be disappointed, for they had succeeded after all in landing on the other bank of the river, and had taken the position there with sixteen guns and a stand of colours. But the Americans were organizing a powerful counter-attack and a thick fog was coming on; Lambert probably did well in getting his men away and being allowed to bring in his dead and wounded. On the 18th he re-embarked and steered for Mobile Bay. On April 8, after some weeks on Dauphin Island, the 43rd and the 7th Fusiliers sailed for England.

In May, when they were off the Land's End, a ship handed them newspapers containing an account of Napoleon's escape from Elba and his successful invasion of France. The 43rd landed, went into quarters at Dover and Deal, received from the 2nd battalion a draft of nine officers and 200 men, and within a fortnight of reaching Spithead were ready again for service. They sailed 1100 strong on June 16, landed at Ostend, and reached Ghent on the 19th to find that the battle of Waterloo had been fought the day before. They then joined the army near St. Denis, and were placed in the 5th division under Sir James Kempt: marched on Paris, and eventually formed part of the Army of Occupation. During their stay in France they were brigaded with the 7th and 23rd Fusiliers under Sir Lowry Cole, and were stationed successively at Melun, Paris, Bapaume, Valenciennes, Cambray and Douchy. Finally, in October 1818, they were at the great review before the Emperor of Russia and the King of Prussia, and marched next day for Calais. On November 1, 1818, they landed at Dover.

1815 In arriving at the front a few hours too late to take part in the most decisive battle of the age, they had suffered a disappointment to which perhaps there is no parallel in the history of the British army. But even if they had been present on the field of Waterloo and had shared the triumph of the 52nd they could hardly have enhanced their reputation. There is a story, current at the time and never called in question, that more than once during the battle, when the Duke's attention was drawn to a dangerous place in the line, he instinctively gave the order "Send the 43rd there!" No battle honours could attest more conclusively the perfection of the regiment's work or the confidence of their chief.

(*a*) (*b*)

BUTTONS OF THE 43RD

(*a*) Worn during Waterloo period and (*b*) after.

CHAPTER XII

1815 (*continued*)

The Waterloo Campaign—Arrival of the 52nd on the field—The battle in outline—The crisis—The part played by the 52nd—Colborne's flank attack on the Imperial Guard—The Duke's final advance with the 52nd—Their attack on the reserve of the Old Guard—Accounts by Eyewitnesses—Value of the regiment's achievement—The march on Paris—A Prætorian cohort—The last to leave France.

Now that the 43rd and the 52nd are completely 1815
united the ill-fortune of the one in so narrowly missing Waterloo may be considered as compensated by the good fortune of the other in being present at the battle by an even narrower chance. The 1st battalion of the 52nd was not only under orders for North America, but had twice embarked and twice been driven back by foul winds, when the news of Napoleon's escape arrived. The regiment immediately sailed again, but this time for the Netherlands: arrived at Ostend on March 31, marched to Brussels on April 4, and on the 7th took over at Grammont the serviceable men of the 2nd battalion, numbering 9 sergeants and 224 rank and file. The 2nd battalion then left for England, and the 1st were cantoned at Queraucamps, near Mons, with Sir Harry Clinton's division.

Late on June 15 Sir John Colborne, colonel of the 52nd, received sudden orders to march. The actual start was made next morning, when Clinton's

1815 division were moved at a moment's notice from their company parades along the Enghien road. From there they heard the cannonade of Quatre Bras at a distance of twenty-two miles, and eventually turned and marched on Brain le Comte, which they reached at midnight in torrents of rain. Two hours later they fell in again, and reached Nivelles about 7 a.m. After four hours' rest they marched again, and about 7 p.m. arrived at Merbe Brain, having been twenty-one hours on the road, wet through and loaded with a double supply of ammunition. They had had their first sight of the French about five o'clock at Mont Plaisir, where some light cavalry were reconnoitring.

Between four and five o'clock on the morning of the 18th, a company of the 52nd and two or three of the 95th were sent to occupy the village of Merbe Braine, which was practically the right flank of the British line. Wellington's position was not actually at Waterloo but more than two miles from there, in front of the tiny village of Mont St. Jean, which stands on the edge of some rolling country with slopes rather like those of the Wiltshire Downs. Mont St. Jean is the meeting-place of two main roads; if Napoleon wished to march to Brussels either from Nivelles or from Charleroi he must come this way, and as there is a long and fairly steep ridge running right across both roads half a mile before Mont St. Jean is reached, the position is a good one for so open a country. Colborne used to say that the Duke had gone over the ground some days before, "and fixed on that place as the one where the battle, he thought, could be fought." He was asked if any entrenchments should be cast up. He said, "No, of course not, that would show them where we mean to fight."

The ridge which formed the front of Wellington's main position runs nearly due east and west; the line of the allied troops was not more than two and a half miles long. The French position was on some irregular ridges opposite; to attack they would have to cross the rolling ground between, and to charge up quite a stiff slope. All along the ridge at the top of the slope there was a road with hedges almost as deep, in parts, as a Devonshire lane, and behind the road was a level plateau, with reverse slopes at the back of it where reserves and hospitals could lie under cover. Napoleon's maps would tell him all about this, but he could not see for himself what was behind the hedge, though he was not far away. The farm of Rossomme, where he sat all day on a high mound, was opposite the centre of the British position, and during the whole battle he was never two miles distant from the Duke himself. 1815

Early on the morning of the 18th he made a careful survey of the British line: he had to decide where to deliver his attack. The Duke's right was too strong: it was covered by a steep ravine near Braine la Leud, and in front of it was the chateau and wood of Hougomont, held by the Guards and some Nassau and Hanoverian troops. Besides, Napoleon's strategical object was to cut Wellington off from the Prussians, who were coming up to join his left. He decided, therefore, to make a feint on Hougomont, and to launch his real attack against the British left; but he was eventually obliged to force the centre too, because there too Wellington had an outwork—the farm of La Haye Sainte by the side of the Charleroi road, occupied by the King's German Legion, with a gravel pit near it, held by the 1st battalion of the 95th.

1815 Napoleon began by a grand spectacular display of his troops, Wellington by riding quietly along his line in his plain blue coat and hat and white buckskins. Soon after eleven o'clock both armies were ready; as the first French gun was heard Captain Diggle, of the 52nd, took out his watch and remarked to Lieutenant Gawler, "There it goes!" It was twenty minutes past eleven. The cannonade opened all down the line, and Jerome Bonaparte at once led the attack on Hougomont with thirteen battalions. They drove in the Nassau troops in the wood and rushed at the hedge of the garden: but behind the hedge was a loopholed wall, and behind the wall were the Guards, who repulsed them with a hot fire. They came on again and again, in front and on the flanks; they even broke into the yard of the farm, but every time they were driven back with the bayonet. The Guards were reinforced by six companies of their own; but no troops were drawn away from the centre, so that the whole move was a failure. Napoleon could wait no longer, for the Prussian advance guard was now visible on his right. He ordered Ney to attack the left centre.

It was now between one and half-past. Ney sent Erlon's infantry up the slope to the left of La Haye Sainte, supported by Kellerman's cavalry and seventy-two guns—18,000 men in all. The infantry were met by Picton with the brigades of Kempt and Pack, who drove them down the slope, and part of the cavalry were routed by a charge of the Household Brigade under Lord Uxbridge. Two columns of French infantry almost enveloped Pack's Highlanders, but then the Royals, the Scots Greys and the Inniskillings charged just in time, and Erlon lost 3000 prisoners, forty guns, two eagles, and an immense number of killed and wounded. Our cavalry, as usual, overshot 1815 and got cut up in their turn, but the French attack was routed.

Napoleon had failed to force his wedge in on the British left; he now determined to take La Haye Sainte and break the centre. He put together a heavy column of Donzelot's and Quiot's divisions, brought up guns from the right and left and attacked the farm on both sides at once. Major Baring, who defended it with his German Legion, fought splendidly and was twice reinforced, but at last his ammunition ran out, the buildings were fired, and what was left of the small garrison had to be withdrawn.

This was a decided success for Napoleon, and one that he had already prepared to follow up at any cost. His infantry had been too much shaken to advance again: there was nothing for it but to throw in the cavalry. Accordingly forty squadrons mounted the slope under a heavy fire and charged when they reached the top. They took the British guns, which in those days were placed in front of the line; but the allied infantry were all formed into squares, against which the cuirassiers could do nothing. They swept between the squares without breaking one of them, and as soon as they were blown and in disorder Lord Uxbridge charged with the heavy brigade and drove them over the edge again. A second attempt failed also, and they retired beaten. Napoleon then threw in thirty-seven squadrons of the Guards cavalry, and the whole mass advanced again. The British line was drawn in to strengthen the centre, the gunners and the staff, with the Duke himself, took refuge inside the squares, and for the third time the flood of horsemen flowed over the plateau like the tide breaking

1815 in among rocks. There was a scene of great confusion, which ended, as before, by the French charge dying away round the unbroken squares and being finally routed by the remnants of the British cavalry. The gunners rushed back to their guns and fired into the cuirassiers as they retreated. Ney had lost one-third of his cavalry, the finest in the world. But he was still game, and he asked for infantry to help him in a fresh attempt. "Infantry?" said Napoleon angrily, "where am I to get them? Do you expect me to make them?" He had already sent all he could spare to hold off the Prussian advance on his right.

But he saw that it was time for a supreme effort. He brought all his artillery into action in a tremendous bombardment which crushed the British batteries; then sent forward all that was left of his infantry line, and finally, when the Allies appeared to be firmly held, he launched his grand reserve, the Imperial Guard. They advanced up the slope against the ridge just above Hougomont, marching in echelon of battalions, which looked from the British position as if they made up two great columns. Ney himself led them on foot—his horse had been shot under him.

This was what was afterwards called "the crisis" of the battle. It was soon over. When the first battalion at the head of the leading column on the right came within distance of Maitland's brigade of Guards, the Duke, who was standing by a small battery beside them, called out, "Up Guards, and make ready." Maitland's men sprang up, fired and charged; the head of the attack broke and fled. At the same time Adam's brigade wheeled down the slope so as to come

on to the flank of the left-hand part of the attack; 1815
they fired into the leading left battalion at close range and then rushed at them. This famous and decisive charge ended the crisis and practically decided the battle; Colborne and the 52nd swept the French Guard before them right across the front of Maitland's brigade, right across the main field of battle, across the Charleroi road and along the other side of it as far as La Belle Alliance, the furthest point reached by any of the Duke's army that day. There the Prussian artillery came in, the whole British line was advancing, and Napoleon rode off. "Tout est perdu," he said, "sauve qui peut," and a complete rout set in. The battle ended with a relentless all-night pursuit by the Prussian cavalry.

For their own part in the battle the 52nd had been waiting nearly all day. Adam's brigade was at first held in reserve; it was not moved forward from its bivouac until three o'clock, when the attack on the centre was beginning. It was then marched to the front and placed on the right centre to relieve the battalions of Brunswick Light Infantry, which had been severely knocked about by artillery fire and cavalry charges. After lining that part of the ridge for some time the brigade was moved about 500 yards down the slope to support the troops defending Hougomont. The 71st were on the right, in battalion square, close to the corner of the Hougomont enclosure; the 52nd were next, in two squares, and on their left rear were the 95th.

This was a useful but uncomfortable position. From a small ridge only 200 yards away a couple of guns and a howitzer fired continuously upon the brigade. "A shell," says Colborne, "came close to

1815 a corner of a column of the 52nd, followed by a ball which passed exactly over the whole column, who instantly bobbed their heads. In the excitement of the moment, more to encourage the men than anything else, I called out, 'For shame! for shame! That must be the 2nd battalion, I am sure.'" (They were recruits.) "In an instant every man's head went as straight as an arrow."

Besides the gunfire, the regiment was threatened by a large body of French cuirassiers who were trying to pass to the rear of Hougomont; every now and again they attempted to charge the 52nd and that was a relief, because then the artillery fire had to stop. After a time the Duke sent Colonel Hervey down with orders for the regiment to withdraw up the hill. Colborne sent an answer back that if it was the danger from the guns the Duke was thinking of, the 52nd could stay where they were, for they were protected by a rise in the ground in front of them. This was not strictly the case, as we have seen, but it was true that only Colborne and his mounted officers were in direct view of the guns.

Half-an-hour after this the Nassau regiment came running in disorder out of Hougomont wood. It seemed likely that Hougomont itself would be taken, and the right flank of the 52nd exposed, so Colborne retired the regiment up to the road on the ridge, and the 71st went back with them. As they went, Colborne, who was riding last under a hot cannonade, heard a Frenchman shouting to him. He turned and saw a colonel galloping towards him—a deserter from the French cuirassiers—and shouting "Vive le Roi!" He came up to Colborne and pointed across the valley to the French centre. "Ce coquin Napoléon est la,"

JOHN COLBORNE, FIELD-MARSHAL LORD SEATON,
LT.-COLONEL OF THE 52ND, 1811-1825.

he said, "avec les Gardes! Voilà l'attaque qui se fait." Colborne put up his glass and saw the Emperor for the first and last time in his life. He was in his great coat, with his hands behind his back, walking backwards and forwards and watching the great columns of infantry marching up to the plateau for the final attack. 1815

A huge mass of the French Guard was coming up towards a point just to the left of the 52nd on the ridge. Colborne became anxious. Maitland's Guards were there behind the hedge, but he could not see them and did not know how the attack was to be met. He had no orders, and no time to ask for any. It was the moment for which he was born and bred, the moment for a bold stroke of his own. He determined instantly to attack the French attack upon the flank.

He could not have done this with any but a perfectly trained regiment, for he meant to place his men, few as they were, in line parallel to the enemy's advance, and it must be done quickly, and without manœuvring under fire. He advanced the regiment down the hill and wheeled it at the same time on its left as a pivot, with one company thrown out to skirmish in front. At that moment the Brigadier, Sir Frederick Adam, rode up and asked what he meant to do. "To make that column feel our fire," said Colborne. Adam rode off and ordered the 71st to make the same move.

There was once a controversy, and there is still some uncertainty, as to which battalions of the Imperial Guard were repulsed by Maitland's brigade of Guards and which were routed and driven across the field by the charge of Adam's brigade, initiated by Colborne and the 52nd. The question, though not of

1815 importance, is of some interest, and several solutions have been proposed; the most reasonable of them would seem to be that the French column, which began its advance in echelon of eight or more battalions, ended by presenting the appearance of two columns owing to the rear battalions having almost overtaken the echelons to their right. But the facts, so far as they concern the 52nd, are not, and cannot be in dispute, for though the Duke's dispatch is unfortunately vague, there is an abundance of evidence from first-rate witnesses to show that he saw and approved Colborne's movement and himself took part in it. Sir Colin Campbell, who was present with the Duke, gives the following account:

"When the column of the Imperial Guard which the 52nd attacked was gaining the summit of the British position, and forcing backward the left company of the 2nd battalion of the 95th, who had become exposed to its fire, the Duke of Wellington, Sir Colin Campbell, and Major Percy were to the right and a little to the rear of Maitland's brigade of Guards. Colborne, then seeing his own left getting into danger, started the 52nd on its 'right shoulder forward' advance. The Duke instantly sent to desire Colborne to continue his movement, and to order the troops on his right to support him."

The Duke then rode with Sir Colin to the 95th, who were on the left of the 52nd and just below him, and called out: "Who commands the 95th?" A voice answered: "I do, sir," and the Duke then called back: "Let the 95th go on." At this moment both Sir Frederick Adam and Lord Hill were sending aides-de-camp to order the 71st to advance on Colborne's right, and Colonel Halkett, still further down the hill, joined in with his Hanoverian battalion. This account 1815 is further borne out by the statement of one of Ney's staff officers, who was present with the Imperial Guard—that although the British troops in front showed " très bonne contenance, nous fûmes principalement repoussés par une attaque de flanc très vive, qui nous écrasa."

The development of the attack is thus described in Moore Smith's *Life of Lord Seaton :* " The company of skirmishers having been ordered to advance without any support except from the battalion and to fire into the French column at any distance, the 52nd, formed in two lines of half companies, after giving three cheers, followed, passing along the front of Maitland's brigade of Guards, who were stationary and not firing. [Sir John Byng, who commanded them, states that their ammunition was expended.] Four companies of the 2nd battalion [of the] 95th were on the left of the 52nd, the 71st and the rest of the division a little behind. As soon as the French column felt the fire of the skirmishing party a considerable part of it halted, and facing to their left towards the 52nd opened a very sharp fire on the skirmishers and on the battalion. The 52nd advanced till they found themselves protected by the hill from the fire of the Imperial Guard. The two right-hand companies having been thrown into some disorder, Colborne called a halt to rectify the line. He then ordered the bugles to sound the advance, and the whole line charged."

At this, says Captain Yonge of the 52nd, " the Imperial Guard, without waiting for the charge, broke, and rushing in confusion obliquely to the rear, involved in their disorder the other troops in echelon to their

1815 right, suffering immense loss from the running fire of the 52nd at point-blank distance. The 71st too, opened fire on the retreating multitude, which to the regiments standing on the higher ground shewed, as it crowded the valley towards La Haye Sainte without a vestige of ranks remaining, like the vast wreck of a great army. Never was disorganisation more sudden or more complete."

The brigade then brought up their left shoulders and pursued. Suddenly a body of cavalry was seen approaching at full gallop. They were at first fired upon, but were seen to be the 23rd dragoons and were allowed to pass through. At this moment the Duke rode up and called out, " Go on, go on." The 52nd went on, but some French guns on the right were still firing into them. Colborne and Rowan both had their horses killed and several other officers were hit, with a good many men. Lieutenant Gawler, of the 52nd, says that he saw the Colonel " suddenly disappear, while his horse, mortally wounded, sank under him. After one or two rounds from the guns he came striding down the front with 'Those guns will destroy the regiment.' 'Shall I drive them in, sir?' 'Do.' 'Right section, left shoulders forward,' was the word at once. So close were we that the guns only fired their loaded charge, and limbering up went hastily to the rear. Reaching the spot on which they had stood, I was clear of the Imperial Guard's smoke, and saw three squares of the Old Guard within four hundred yards further on . . . standing in perfect order and steadiness. . . . Colonel Colborne then called the covering sergeants to the front and dressed the line upon them. The Duke and Lord Anglesea and his aide-de-camp came up in

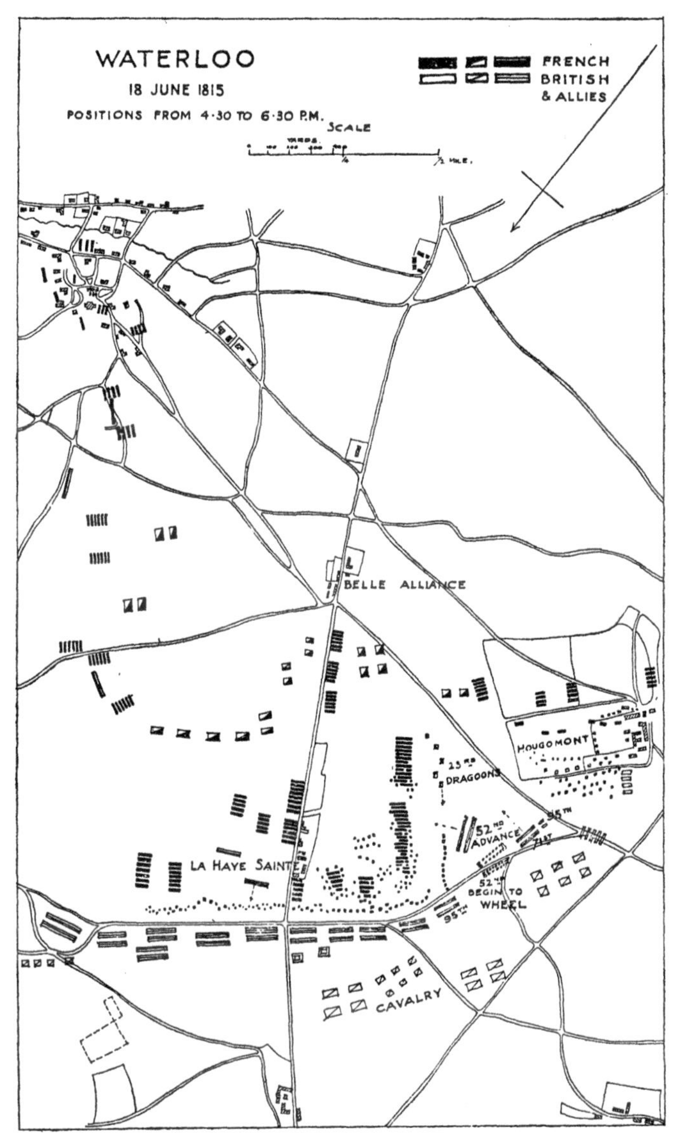
WATERLOO
18 JUNE 1815
POSITIONS FROM 4·30 TO 6·30 P.M.
SCALE
FRENCH
BRITISH
& ALLIES
BELLE ALLIANCE
HOUGOMONT
23RD
DRAGOONS
52ND
ADVANCE
95TH
71ST
LA HAYE SAINTE
52ND
BEGIN TO
WHEEL
95TH
CAVALRY

1815 rear of the 52nd centre. . . . Up to this moment neither the guns, the squares of the Imperial Guard, nor the 52nd had fired a shot. I then saw one or two of the guns slewed round in the direction of my company and fired, but their grape went over our heads. We opened our fire and advanced: the squares replied to it and then steadily facing about retired. . . . We then made a sweep on the top ridge of the ground to get towards Rossomme; twilight had manifestly commenced, and objects were now bewildering. The first event of interest was, that getting among some French tumbrils with the horses attached, our Colonel was seen upon one shouting: 'Cut me out.' Then came some long shots from the Prussian guns far away on our left; still the square of the Imperial Guard was retreating in order, and within 250 paces of my company. Then we came upon the hollow road beyond La Belle Alliance, filled with artillery and broken infantry. Here was instantly a wild *mêlée*. . . . The next event was a French gun, about thirty yards to our right, wheeled round by Campbell (our brigadier's aide-de-camp), and some 71st men, and discharged at our Imperial Guard Square, then that square on the flat of the hill, about a quarter of a mile short of Rossomme, halted, threw off their knapsacks, and again went off in order. We passed these packs lying in square, and soon afterwards halted with our leading company about 100 yards south-east of the south-east corner of the wall of Rossomme . . . The Duke was behind us soon after we halted; it was then so dark we could only discern figures. . . The Prussian regiments as they came up the road from Planchenoit and wheeled round into the great chaussee of Rossomme, moved in slow

time, their bands playing our national anthem, in 1815
compliment of our success; and a mounted officer at the head of them embraced the 52nd regimental colour to serve as the expression of his tribute of admiration for the British Army."

On the scope and value of the 52nd's achievement it is perhaps worth while to cite one authority. General Sir James Shaw Kennedy, of the 43rd, who was Quartermaster-General to the 3rd division at Waterloo, wrote some years afterwards as follows: "I don't think that I impose upon myself a formidable task

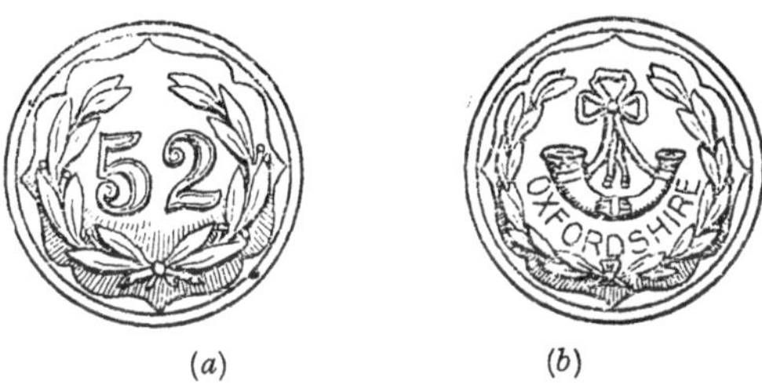

BUTTONS OF THE 52ND.
(*a*) Worn previous to 1882; (*b*) since 1882 (officers').

when I say that no man can point out to me any instance, either in ancient or modern history, of a single battalion so influencing the result of a great general action as the result of the battle of Waterloo was influenced by the attack of the 52nd regiment on the Imperial Guard, of which it defeated first four battalions and afterwards three other battalions; and Colborne did almost all this from his own impulse and on his own responsibility."

The 52nd marched south next day and reached Paris on July 4. The British army was encamped in the Bois de Boulogne: Adam's brigade alone were

1815 selected for the honour of occupying the city. The three regiments marched in on July 7, by the Barrière de l'Étoile, and were stationed in the Champs Elysées; two companies and the quarter guard of the 52nd by the garden wall of the Duke of Wellington's house—perhaps the proudest position ever held by a *corps d'élite.* The regiment was afterwards quartered at Racquingham and Valenciennes until the great review of October 23, 1818. On November 23 in that year it embarked at Calais—the last of Wellington's army to leave the soil of France.

CHAPTER XIII

1819–1859

The 43rd and the Shorncliffe Tradition—The Canadian Revolt—The Great Winter March—The Kaffir War—The Wreck of the *Birkenhead*—The 52nd and the Shorncliffe Tradition—The 43rd and 52nd in the Indian Mutiny—The March of the 43rd through Central India—The 52nd at Delhi—The storming of the Cashmere Gate.

WE have now entered upon that long period—extending over exactly ninety-nine years—during which the British Empire was at peace, except for a single interval of three years, with all the civilized powers. The single interval was the Russian War of 1854–7, in which, as Lord Salisbury said long afterwards, "we put our money on the wrong horse." Our contingent in the Crimean Expeditionary Force, though small and ill-handled, did enough hard fighting to cover the mistakes of their commanders; but the war was not favourable to the display of true military efficiency, and the 43rd and 52nd probably missed no great opportunities by their absence from the field of operations. In almost all the other Imperial wars and expeditions of this period one or both of the two regiments were engaged, and their work throughout sustained their reputation—an arduous and proud achievement after such a past.

From 1819 to 1823 the 43rd were in Ireland; from 1819

1825 1823 to 1830 at Gibraltar. In 1825 they heard a cheering echo of the old Shorncliffe rhyme:

> "No cavalry in England can form a line so quick
> As the 43rd Light Infantry can at the double-quick!"

The occasion was a field day carried out before General Foissac la Tour, then commanding the French army in Spain. At the end of the manœuvres the General said to Major Booth, who commanded in the absence of Colonel Haverfield, "This day has disabused me of an error of twenty years. I always thought the French infantry the quickest to move in Europe, but they are nothing to you, you move like cavalry!" The words were one more salute over the grave of Sir John Moore.

After some time at home the regiment sailed again for Canada, and remained there for ten years, during
1837 two of which, 1837–8, they played a characteristic part in the suppression of the Canadian Revolt, or Tapineau's Rebellion. The outbreak had taken place in the Lower Province, and was promptly dealt with by Sir John Colborne and the troops at Montreal; but there was great anxiety lest the movement should spread to Upper Canada. The Governor of New Brunswick offered the services of the 43rd, then quartered at Fredericton and St. John's. The offer was accepted, and Colonel Booth, with two companies of the regiment, set out upon an expedition unique in the history of the British army. Their object was not to crush rebellion, but to forestall it; to discourage the disaffected and wavering by a display of resolution and efficiency. For this purpose they were to march in the depth of winter across the Portage of the Mada-

waska to Quebec—370 miles of forest and mountain, 1837
in a temperature fifty to sixty degrees below freezing-point. They left Fredericton on December 11 and covered the distance in eighteen days. They marched by Woodstock, Tobique, Grand Falls, thence on the ice to the Madawaska Settlement, over to the Temiscouata Lake, along the right shore and across the famous Portage—thirty-six miles of mountain track—to the south bank of the St. Lawrence near the Rivière du Loup, and finally for 110 miles along the riverside to Pointe Levi, where the shades of Wolfe and Monckton must have watched them as they embarked for the crossing to Quebec.

Their reception was a memorable one: for a month all Canada had been talking of them, and here at Quebec the quays and wharves were crowded to see them land. Through a lane of soldiers and cheering townsmen they marched to their quarters in an old Jesuit convent—"the ragged, unshaven, smoke-dried, toil-worn, frost-bitten 43rd"—but a 43rd worthy of the old Light Division. Happily the Duke was still alive to appraise their work. For him it was "one of the greatest feats ever performed, and the only military achievement by a British officer that he really envied."

The regiment left Canada in 1846 and was stationed
in the south of England and in Ireland until 1851. In 1851
October of that year it sailed for the Cape, and joined the field force in British Kaffraria under Sir Harry Smith, who was delighted to see the 43rd again. The campaign was a trying one, and lasted through a whole year; it was entirely successful, and might long ago have been forgotten among our many little "native wars" but for one immortal triumph of discipline.
On February 26, 1852, H.M.S. *Birkenhead*, carrying 1852

1852 drafts for the regiments at the Cape, was on her passage from Simon's Bay to East London. Among the 700 aboard her were a sergeant and forty privates of the 43rd, under Lieutenant Girardot, to whom, with Captain Wright of the 91st, belongs the chief glory of this heroic tale.

In a calm sea, under a starlight night, the ship was going eight knots when she struck a sharp rock off Point Danger. The Captain immediately let go an anchor and got out the boats; unfortunately he also gave the order to go astern. The ship backed off the rock and began to settle down by the head; the engine-rooms were swamped, and many soldiers were drowned in their berths. Colonel Seton of the 7th and Lieutenant Girardot took their men to the pumps, where they worked in perfect order until the horses had been thrown overboard and the women and children put into the boats, which then stood off about a hundred yards from the wreck.

It was now twenty minutes since the first shock, and the ship was breaking up. The fore part went down suddenly, and the survivors were crowded upon the stern, which was sinking more gradually. The Captain, seeing that the end was come, gave the order, "All those who can swim, jump overboard and make for the boats." But he was not obeyed. The troops had already given all the available boats to the women and children; they remained themselves in unmoved order, "far exceeding," said one who saw them, "anything that I thought could be effected by the most perfect discipline. Everyone did as he was directed, and there was not a murmur or a cry amongst them until the vessel made her final plunge. All received and carried out their orders as if embarking for a

world's port in lieu of eternity. There was only this 1852
difference, that I never saw any embarkation conducted with so little confusion."

Of these men one more thing was now demanded—the sacrifice of their last desperate hope of safety. When the order was heard that they should make for the boats, the voices of Wright and Girardot were also heard begging them to remember that the boats were already full, and that if they were swamped by a crowd of swimmers the women and children must all be lost. It might be thought that in face of a positive order such an appeal at such a moment could only succeed by a miracle. With man all things are possible: the miracle occurred, the ranks remained unbroken, and the ship went down under a company as truly undefeated as any who ever died in battle.

Not all were lost; a few escaped the sharks and the impenetrable net of seaweed which fringed the coast. Among them were Wright and Girardot, and Cornet Bond of the 12th Lancers: three heroic swimmers, and Girardot the strongest of the three, for after ten hours in the water he made his way to shore carrying with him from the wreckage a man who could not swim. A month later he reached King William's Town with the fourteen survivors of his detachment. Sir William Napier brought his doubly distinguished service before the Duke of Wellington; but in those days crosses did not exist, and the rules absolutely forbade the promotion of a subaltern for merit. This was a loss to Girardot's contemporaries, for no generation can afford to neglect such officers; and a loss to Girardot, too, for the remainder of his transient career. But the essential man, the man who in the last extremity could command the animal not only in himself but in others, gained that which is more than all possible rewards. The wreck of the *Birkenhead* has become one of the world's stories; to this day it is told for an example even among our enemies.

Two years later the 43rd left South Africa for India, a part of the Empire where the regiment had never yet served in the 113 years since its first enrolment. They
1854 landed at Madras on January 30, 1854. The 52nd were now also in India, having landed at Calcutta only two months before, and both regiments, though not employed in the Crimean War, were at hand for the infinitely more important service of crushing the Sepoy Mutiny.

The 52nd had seen much foreign service, but no
1823 fighting since Waterloo. From 1823 to 1831 they were stationed successively at St. John's, Fredericton and Halifax, Nova Scotia; then, after two years at home, they spent two more at Gibraltar and three in the West Indies. In 1842 they returned to St. John's and Fredericton; from 1844 to 1847 they were at Quebec and Montreal; from 1847 to 1853 at home again. That they, no less than the 43rd, had faithfully kept their tradition through many years of peace is proved again and again by the General Orders of the Commanders who reviewed them. The following is a typical comment taken from the Orders of General
1839 Sir Samford Whittingham at Barbados, in 1839:

"The 52nd Regiment is, and has long been, one of the most brilliant corps of Light Infantry in the British Army, and its discipline in the field is equalled by its good conduct in quarters.

"It is now thirty-five years since the ever-to-be-lamented Sir John Moore undertook the organization

of the 52nd as a Light Infantry battalion. What com- 1839
plete success has attended his efforts the whole British Army can testify. The British Light Infantry is now second to none, and the 52nd Regiment is a beautiful specimen of the hand which formed it; but to the admirable interior system adopted by Sir John Moore, the durability of the superior discipline of the regiment must be attributed."

The General then summarizes this "interior system"—the groundwork of it is the elementary drill, but the part of it to which he wishes to call the attention of the other regiments under his command is the complete responsibility of the Captains for their own companies and the extent of the power delegated to them for this purpose. It is this system, he says, which, invariably acted upon, "has preserved the 52nd for thirty-five years in its present splendid condition." The result was good conduct in quarters, and in the field great "activity and precision"—he instances a very long advance in line and a subsequent charge which could not have been better executed.

It is hardly necessary to remark that for the troops 1857
engaged in fighting the Indian mutineers these qualities of activity and precision were two vitally necessary ones. The problem was not how to meet phalanx with phalanx in orthodox manœuvres, but how to cover immense distances by forced marches and to deliver attacks with perfect steadiness and small numbers against well-armed but only half-disciplined hordes of much greater strength.

Of the two regiments the 43rd distinguished itself chiefly in the work of the flying columns, the 52nd in the actual fighting. The 43rd at this time was in perfect

1857 condition: its strength was thirty officers and 969 men, and it is stated to have been "entirely composed of the finest description of seasoned soldiers ever remembered by the most veteran officers." These men in their great march through Central India at the hottest season of the year may be said to have broken all the marching records of the British army; they covered 1300 miles in all, and during the first stage of 950 miles, when they were on their way to join Brigadier Whitelock's force for the attack of Kirwee, the total loss of the regiment was only two deaths, both accidental. At Kirwee, where the brigade captured 42 lacs of rupees and an enormous treasure of gold and jewels, three companies of the 43rd were left as garrison. The remaining five companies continued the march to Calpee, and suffered severely from sunstroke and other heat diseases. During the latter half of 1858 and the whole of 1859 detachments from the brigade were continually engaged in breaking up rebel forces in the hills and jungles of the Jumna country. A rebel attack on the Kirwee garrison was beaten off; at Purwanee a force was defeated in the open and four guns taken; and in a fight in the Punnah jungle Private Addison won for the regiment its first Victoria Cross by rescuing Captain Osborn, the political agent, who had been cut down by a Sepoy.

In the meantime the 52nd also had been distinguishing itself, and it had the good fortune to be nearer to the chief centre of operations. When the Mutiny broke out the regiment was at Sealkote, and at once prepared to join Brigadier-General Neville Chamberlain's column at Wuzurabad. They marched on May 25 in light order as rear regiment of a small column, and within two hours were overtaken by a terrific dust-storm. Their dogged resolution and the extraordinary violence of the storm are both proved by a curious incident. The 52nd resumed their advance at the earliest possible moment, and without knowing it marched in the darkness of the storm right over the 35th Native Infantry, who lay in the road among their transport animals, still buried in dust. 1857

On July 10 the regiment made a forced march against the rebels at Goordasepore—42 miles in less than twenty hours. On the 12th they found the enemy, 800 infantry and 300 cavalry, and though weary from their march they " pressed on as if fatigue was unknown to them." The fight was a strange one to the eye, for the rebels were all in British uniform, with British colours, while the British were for the first time in *karkee*, which had been procured by Colonel Campbell before leaving Sealkote. The 52nd had the advantage of being armed with the new Enfield rifle, then considered " the very king of weapons," and their enthusiasm was great, for the regiment was going into action for the first time since Waterloo, and they longed to show that forty-two years of peace had left no rust upon their steel. Their first charge was decisive; in two days' fighting they destroyed a force three times their own number and broke the rebellion in the Punjab.

On August 14 the 52nd joined the British troops before Delhi. The regiment marched in 680 strong, with only sixteen sick, but by September 13 it could only muster 240 fit for service. They were enough. At midnight on that date the breaches in the curtain wall between the Water Bastion and the Cashmere Bastion were reported practicable, and the assault was ordered to be delivered at once. It was to be made by

1857 three columns, and all three immortalized themselves: to the 52nd fell the honour of a principal share in the most memorable attack of all.

The first column, made up from the 1st Bengal Fusiliers, the 75th Foot, and Green's Punjabis, and commanded by John Nicholson, was to carry the main breach; the second, consisting of the 84th Foot, the 2nd Bengal Fusiliers and Rothery's Sikhs, was to storm the breach in the Water Bastion; the third, which contained the 52nd, the Kumaon battalion and the 1st Punjab Infantry, was commanded by Colonel Campbell of the 52nd, and was to rush the Cashmere Gate as soon as it had been blown in by the engineers.

By daybreak all were in position; soon after sunrise the British guns ceased firing and Nicholson gave the signal. The first and second columns rushed to the breaches; the third stood still to witness one of the most heroic deeds of our history. Lieutenants Deacon, Charles Home and Philip Salkeld, of the Engineers, and three Sapper sergeants, Smith, Carmichael and Burgess, advanced to the gate; with them went Bugler Hawthorne, of the 52nd, to sound the advance when the work was done; behind them eight native sappers under Havildar Madhu carried the bags of powder to be piled against the gate. Under the glacis, in front of the waiting column, lay Captain Bayley commanding the forlorn hope company of the 52nd; fifty yards behind him were the supports, consisting of fifty men of the 52nd and fifty of each of the two native regiments.

Of the seven against the gate five fell at once; but Salkeld, though terribly wounded, fastened the powder bags in place and laid the hose; the light he handed as he lay to Sergeant John Smith, who fired the train, and Bugler Hawthorne sounded the advance. The stormers rushed forward irresistibly; Captain Bayley 1857 fell before he reached the gate, but Captain Crosse, of the 52nd, was close up with the supports. He led through the gate, and the entrance was won without a check. Colonel Campbell and his Brigade-Major Synge were among the first six; the column was reformed inside the main guard, cleared the Water Bastion with the bayonet, cleared the adjoining compound and houses, the church and the Cashmere Durwaza Bazaar. A gun commanding the street was taken by a charge in which Lieutenant Bradshaw of the 52nd was killed; the gate of the Duruba was burst open, and the Jumma Musjid came in sight. There the attack was stayed for lack of ammunition or explosives, and Colonel Campbell retired to take up his position in the church, where the 52nd remained that night. During the fierce days of fighting which followed they advanced with the 60th Rifles from one strong post to another, till on September 20 the palace was taken and the enemy fled from the whole city.

The fame of the Cashmere Gate resounded even above the din of those Homeric days. Of the seven against the gate the five survivors all received the Victoria Cross: the three officers and Sergeant John Smith for conspicuous gallantry in the performance of their desperate duty; and Bugler Robert Hawthorne for that he not only most bravely performed the dangerous duty on which he was employed, but also bound up Lieutenant Salkeld's wounds under a heavy fire, and had him removed without further injury. A sixth Victoria Cross was given to Lance-Corporal Henry Smith, and a commission to Sergeant-Major Streets, both of the 52nd, for their conduct during the fighting in the city. " At the siege and assault of Delhi," said

1857 Colonel Campbell in his Regimental Orders, "the conduct of the regiment fully realized the most ardent expectations of its commanding officer, and it is with the greatest joy and pride that he thus testifies to its admirable behaviour." In this rare departure from the laconic tone of the profession, George Campbell was well inspired: he spoke out not for himself only, but for every officer who, in time of active service, has had the fortune to command the 52nd.

CHAPTER XIV

1859–1914

The 43rd in New Zealand—At the Gate Pah and Te Ranga—The 43rd and 52nd linked as the Oxfordshire Light Infantry—Foreign service of the two battalions—The 52nd in the Mohmand and Tirah Campaigns—The 43rd in the Boer War.

THE 52nd returned to England in June 1859; the 43rd 1859
remained in India until September 1863, when they were ordered to New Zealand, where the Maories were giving trouble. The campaign lasted two years and was one of extreme difficulty, waged by small forces against an enemy of great courage and intelligence, fighting in defence of his own country. The two principal engagements were the assaults of the stockaded redoubt known as "the Gate Pah," and the entrenched position of Te Ranga. In both of these the regiment took a principal part: the second was a complete success, the first a defeat, redeemed only by the magnificent courage of the officers of the 43rd, who led the storming column.

The force assembled in front of the Gate Pah on the
evening of April 28, 1864, consisted of the 68th Regi- 1864
ment, 700 strong, about 300 of all ranks of the 43rd, a Naval Brigade of 400, a Movable Column of 180, and a detachment of Artillery with seven guns and eight mortars. The 68th made a flanking movement after dark

1864 and succeeded in getting to the rear of the position by passing along the beach at low water. At daybreak the artillery opened on the stockade, which was immensely strong; it was not until four o'clock in the afternoon that a breach appeared practicable.

The assaulting column consisted of 150 of the Naval Brigade under Commander Hay of the *Harrier*, and 150 of the 43rd, with their Lieut.-Colonel, H. J. P. Booth, who was to lead the attack. The two detachments charged simultaneously and in brilliant style; they crossed the open ground with small loss and carried the breach. But inside the Maoris met them with the greatest fierceness, and a desperate struggle followed. Colonel Booth and Commander Hay, with Captain Glover and Ensign Langlands of the 43rd, were first into the fort; all four fell dead or mortally wounded, and in a few moments almost every officer in the column was down. The enemy were firing from all sides, darkness was coming on, the men were leaderless and confused by the intricacy of the inner defences. Some sailors shouted the warning, "They are coming into us in thousands!" and the brigade retired to take cover outside the position. The remnant of the 43rd followed, leaving nine officers and thirty-two of their men dead or wounded in the fort.

General Cameron was perhaps wise in not throwing in his reserve, which only numbered 300 men of the 43rd and the Marines. He entrenched for the night within 100 yards of the Pah and sent to his base for reinforcements. But the heroes of the assault had not fallen in vain. The Maories had suffered severely in the fighting and could not face a second attack; in the darkness of the night they evacuated the position and crept away through the lines of the 68th.

For the 43rd this day remains a proud and bitter memory. In the charge the men of the small force which represented the regiment showed all the old swiftness of attack: in the moment of confusion two-thirds of them fell short of the old steadiness. But the officers and those who rallied round them were beyond praise: none ever did better, even in the annals of the Light Infantry. In the enemy's position, dead or dying, lay the Colonel with his four captains, Glover, Mure, Hamilton, and Utterton, beside them Lieutenant Glover and Ensign Langlands. Of the four remaining officers, Lieutenant and Adjutant Garland and Ensigns Nicholl, Clark and Garland, two were wounded, and all received the thanks of the Commander-in-Chief. In the same honour were included Sergeant Garland, Corporal Harrison and Privates Bridgeman and Maitland; and Lieutenant Garland was specially promoted without purchase. The names of these men may well be remembered in the 43rd. 1864

Shoulder Badge worn during Queen Victoria's Reign.

The enemy's position at Te Ranga was very similar to that of the Gate Pah, but the attack upon it ended very differently. A body of some 600 Maories were entrenched in a chain of rifle-pits across the road four miles beyond their former stronghold, and on June 21 Colonel Greer advanced against them with a force of about the same number, including ten officers and 230 men of the 43rd. Skirmishers were driven in by a line of the 43rd and 68th, under Major Synge, and a hot fire was kept up for two hours until the supports could be brought from the camp. As soon as they arrived the

1864 advance was sounded and the two regiments charged together. Major Synge's horse was shot under him; Captain F. A. Smith was first into the rifle-pits on the right, Major Colville among the first on the left. The enemy, as before, made a fine fight for it. "They stood the charge," says Colonel Greer in his dispatch, "without flinching, and did not retire until forced out at the point of the bayonet, which everywhere did its work."

Six officers of the 43rd were mentioned for good work in this affair, and Captain F. A. Smith, who was very severely wounded, received the Victoria Cross. General Sir Duncan Cameron in forwarding the dispatch reported that "the valour and discipline of the troops . . . were conspicuously displayed on this occasion, and the 43rd and 68th Light Infantry, on whom the brunt of the engagement fell, behaved in a manner worthy of the reputation of these distinguished regiments." In the following March Major Colville was promoted to Brevet-Lieut.-Colonel and Captain F. A. Smith to Brevet-Major for distinguished service at Te Ranga.

The regiment returned to England at the conclusion of the war in February 1866. Seven years
1873 later, when it was again in India, a detachment stationed at Malliapoorum was called out against a band of Moplah fanatics. The officers in command, Captain Vesey and Lieutenant Williamson, received the thanks of the Government of Madras for their energy in quelling the disturbance on the first day of its existence.

1881 In 1881 the regimental system of the British army was recast: the 43rd and 52nd were united as 1st and 2nd battalions of the Oxfordshire Light Infantry, and attached to the 43rd Regimental District with Depot Headquarters at Oxford. Probably in no other case

was the linking together of the hitherto independent 1881
regiments so clearly suggested by common traditions and a common fame. The 43rd and 52nd had served more often together than apart: in a period of one hundred and six years they had fought side by side in three great sieges and more than a dozen pitched battles: under Moore they had together set the pattern of the Light Infantry; under Craufurd and Alten they had together perfected it in seven years of war. Their old and close comrades, the 95th, had long ago been separately established as the Rifle Brigade; the name and system of the Light Division could not be more fittingly perpetuated than by the union of its two Light Infantry Battalions.

It may be added that the loss of individuality, so freely and even bitterly prophesied as a result of the new system, has turned out to be an imaginary danger. The territorial idea in 1881 was no new one, and while it has proved more useful since that day than before, it has neither destroyed the distinguishing characteristics nor superseded the historic numbers by which our old and famous regiments were known. The 43rd and the 52nd have in no degree merged or surrendered their individuality. Like brothers they bear the same name and inherit the same virtues; but like brothers they keep, and perhaps even accentuate, the differences of custom and dress which are among the marks of vigorous life.

In December 1884, as if to prove that no breach of 1884
continuity had occurred, the Moplah fanatics broke out again, and were again suppressed by the 43rd—the first active employment of the regiment as a "1st battalion." For this service the officers and men engaged were honourably mentioned in the Orders of the Government

of India, and Private Barrett was awarded a medal for distinguished conduct.

1885 In 1885 the 52nd began their active service as a "2nd battalion" by sending a detachment under Lieutenant Scott to act as mounted infantry in the Nile Expedition. In 1890 the battalion, then stationed at

HELMET PLATE WORN DURING QUEEN VICTORIA'S REIGN.

Burma, provided a company as escort to the Siamese Boundary Commission, and in the following year two more companies formed part of a column operating in the Wuntho district in Upper Burma. In 1897 the 52nd was engaged in the Mohmand campaign on the North-West Frontier of India, and was present on September 27 at the successful action at Koda-Khel. On the conclusion of this expedition the battalion was immediately ordered to form part of the Peshawar column of the Tirah Expeditionary Force, under Brigadier-General Hammond.

The campaign of 1897–8 was probably the most 1897
difficult which had been undertaken by a British army since the first Afghan War. The country to be traversed is described by Colonel Hutchinson as one of " high mountains, precipitous cliffs, dangerous defiles, wild ravines, rushing torrents everywhere, while roads of any kind were conspicuous by their absence . . . marches must be performed under conditions in which they become slow, exhausting processional movements, the long trailing flanks of which are exposed to attack from start to finish." The enemy too were formidable in themselves—mountaineers of a fine type, bred amongst the hills, inured to the climate with its deadly alternations of extreme heat and cold, and accustomed from childhood to the use of arms and the practice of every kind of treachery: " hardy, alert, self-reliant and active, full of resource, keen as hawks and cruel as leopards." They were at this date well armed and formidable marksmen; many of them had served their time in the British service and were familiar with the strong and weak points in our organization—translations of English drill books and musketry regulations were found in some of their deserted houses. " The boast of the tribes," said General Sir William Lockhart, the Commander-in-Chief of the Expedition, in his farewell Orders, " was that no foreign army, Moghul, Afghan, Persian or British, had ever penetrated, or could ever penetrate, their country; but after carrying three strong positions and being for weeks subsequently engaged in daily skirmishes, the troops succeeded in visiting every portion of the Tirah, a fact which will be

1897 kept alive in the minds of future generations by ruined forts and towers in their remotest valleys."

In this arduous campaign the 52nd were more actively employed during the later than the earlier part of the operations. Upon the evacuation of the Tirah country the Afridis had failed to comply fully with the terms imposed by Government, and Sir William Lockhart ordered an immediate advance to be made against their other winter settlements. Accordingly on December 15 the Peshawar column marched from Swaikot for Jamrud, was joined there by the 1st division and the Gurkha Scouts, made a reconnaissance into the Khyber, and began the work of repairing the road. On the 23rd the fort of Ali Musjid was reached and the towers of the village of Lala China were blown up. Sir William Lockhart himself joined at this point, bringing with him General Sir Henry Havelock-Allan, V.C., M.P., who had come from England as the Commander-in-Chief's guest.

On December 26, while the 1st Division explored the Bazar Valley, General Hammond's column marched to Landi Kotal, where they found the *serai* partially destroyed and the quarters looted. Here the 52nd were set to work repairing the fort, demolishing village defences of the Zakka-Khels all along the line of the Khyber, picketing the hills which command the route, and convoying supplies sent up from the base. Here, too, they had their sharpest brush with the enemy. On December 30 their forward picket began to retire at 3.30 in the afternoon and had joined with two other sections near a Buddhist *tope*, when a volley was fired into them from their left rear. Three men fell, and the sections got under cover in a nullah. Two of the wounded, who could walk, were sent along the nullah

to the surgeon; the third was dressed on the spot by 1897
Colonel Plowden. The retirement was then continued, but in crossing an open gap Corporal Bell was killed by a shot in the head. Colonel Plowden, with Lieutenants Fielden and Owen, carried him some way up the nullah, but at another break in the bank Private Butler was shot in the leg, and a halt was made while Captain Parr and Lieutenant Carter dressed his wound. Lieutenant Carter then hoisted him on his back and set out to carry him across the open space. When they were half-way across the wounded man was again hit and killed; Lieutenant Carter fell under him from the force of the blow. Lieutenant Fielden then dashed across, and the two officers carried the dead man under cover. Immediately afterwards Colonel Plowden and Lieutenant Owen were both wounded while carrying Corporal Bell's body across the same place.

In the meantime, Bugler Crowhurst of the 52nd was galloping into Landi Kotal on Colonel Plowden's charger with a note asking for support. General Hammond started out at 5.30 and brought off the whole force in safety, with the dead and wounded whom they had saved from the Afridi knives. A two or three hours' skirmish is a small incident in a campaign, but this one deserves to be recorded because it is typical of a frontier war, and, from a regimental point of view, important as a proof of fighting quality. That the Afridis were in earnest is clear from their simultaneous attack on the Ali Musjid pickets; and it was on this same afternoon that they killed General Sir Henry Havelock-Allan on his way down to Jamrud. But in the three first
months of 1898 they realized the hopelessness of their 1898
struggle, and paid in the fines and rifles demanded of them. By April 5 they were crowding in hundreds

1898 round Sir William Lockhart's house in the cantonments at Peshawar, cheering him enthusiastically as he left for England. A more curious scene the 52nd have probably never beheld.

The 43rd were now to have their turn of active service. They had returned from India at the beginning of 1887, after transferring 310 men to the 2nd battalion; the deficiency was made good on their arrival at Shorncliffe by the addition of 400 men who had been recruited in readiness as a Provisional Battalion. After this, however, recruiting was checked by the authorities, with the result that when the South
1899 African War broke out in October 1899 the regimental reserve was only able to furnish 361 men, and the battalion was not included in the 1st Army Corps. On December 2, however, it was mobilized with the 2nd Buffs and 1st West Riding Regiment (the 2nd Gloucesters were afterwards added) as the 13th Brigade under Major-General Knox, and formed part of the 6th Division, commanded by Lieut.-General Kelly-Kenny. On December 11 the regimental colours were deposited with great solemnity in Christ Church Cathedral, Oxford, and on the 22nd the battalion embarked at Southampton.

1900 It reached Cape Town on January 14, a few days after Lord Roberts had landed. At that moment all the British forces were held up by the enemy—Buller in front of Colenso, Methuen at Magersfontein, French in Cape Colony—but the tide was on the point of turning. On February 11 French set out on his great cavalry movement; on the 15th Kimberley was relieved and Cronje left his trenches to make a dash for Bloemfontein. Kelly-Kenny pursued him closely, and the 43rd were engaged all day in the action with his rearguard at Klip Kraal, losing fifty men killed and wounded. They made little progress this day, but on the 16th the brigade captured part of the Boer transport, including eighty waggons of stores, rifles, cartridges and shells. Two days later they came up with Cronje's army at Paardeberg and took part in the attack. The 43rd were in action from 8 a.m. till dark, and fired 17,260 rounds of ammunition; their losses were three officers killed and four wounded, six men killed and twenty-six wounded. On the 27th Cronje surrendered with his whole surviving force, 4072 in number. 1900

On March 1 the brigade heard the news of the relief of Ladysmith. On the 13th they marched into Bloemfontein, but left again on the 31st for Krantz Kraal Bridge and other positions defending the town on the south-east. On April 11 the 43rd were reinforced by a draft of five officers and 300 men from England, and a second draft of two officers and 100 men followed on the 16th. On May 8 the strength was brought up to 900 by the arrival from England of a Volunteer Company enlisted for one year or the duration of the war. In the ranks were "a lot of Oxford Undergraduates." The brigade was now broken up and the regiments were separately employed, at first on the lines of communication, afterwards in the long chase after De Wet, in which, during the latter part of the year, the 43rd marched upwards of 700 miles. The years 1901 and 1902 were occupied in the pursuit of the Boers in the Orange River Colony, in the numerous great "drives," and in manning the blockhouse lines.

The Mounted Infantry Company of the 43rd also did good service in this war. They sailed three weeks later than the regiment, reached Cape Town on Febru-

1900 ary 8, and were sent up to join General Clements' force at Arundel. Their service during the next nine months was of the most active kind, but they were not heavily engaged with the enemy until November 5, when they took part in the action at Bothaville, which General Knox afterwards described as "one of the hottest fights of the war." De Wet and Steyn were laagered at Bothaville with eight guns and 2000 men, and a flying column under Colonel Le Gallais surprised them by a rapid march, the advance guard getting in touch with them and capturing their pickets at daybreak on the 5th. Captain Colville, with one half of the Mounted Infantry Company of the 43rd, at once seized a farm on the right 200 yards from the Boer laager, and then, supported by two guns and the 5th and 7th Mounted Infantry, he held on under shell and rifle fire for four hours. The other half of the company acted as escort to two guns of "U" Battery, and had a desperate fight for three hours with some unusually determined Boers who were trying to work round the left flank. Ammunition began to run short, and the situation was serious; but at 7.30 reinforcements came in, followed soon afterwards by Colonel De Lisle with two companies of Australians, who had galloped up from twelve miles away directly they heard the firing. By 9 a.m. the Boers had retired on both flanks, and the 150 who remained were surrounded in the farm and enclosure, where they had two of their guns, the remainder being out of action in the laager. It was arranged to give them three minutes of magazine fire and then charge with the bayonet, but after one minute's firing they hoisted the white flag. Seventeen wounded and ninety-seven unwounded Boers surrendered, with seven guns; and Captain Colville's company was

LORD RAGLAN, CAPTAIN IN THE 43RD, 1808, AND MILITARY SECRETARY TO LORD WELLINGTON.

specially mentioned in the commanding officer's report 1900
for their share in a finely fought action.

The losses of the regiment during the war were three officers killed and eight wounded, twenty-eight men killed, eighty-seven wounded, and eighty-four died of disease, a total of 210 out of the thirty-three officers and 1350 men who from first to last were landed in South Africa. Of the officers, seven were mentioned in dispatches: Lieut.-Colonel Hon. A. E. Dalzell received the C.B., Major Fanshawe and Captain Ruck-Keene the D.S.O., and Major Hanbury Williams was promoted to Lieutenant-Colonel. These are honourable facts, but there are others equally eloquent of the spirit of the 43rd. Of the eight officers who were wounded (most of them severely), six returned to active service in the war within a few months. Of the men two were given commissions, and fifteen received the medal for Distinguished Conduct.

The 52nd in the Great War

CHAPTER XV

1914

The War of 1914—The British Expeditionary Force—The 52nd at Mons—Rearguard in the Great Retreat—The battle of the Marne—The fighting on the Aisne—The Cour de Soupir—The rush to the North—The 52nd at Langemarck—With the 7th Division at Zandwoorde—The 31st of October—Relieving the 4th Brigade—The Model Trenches—The crisis of November 11—The defeat of the Prussian Guards.

1914 On the 1st, 2nd, 3rd and 4th of August, 1914, the storm of war broke in successive gusts upon the peace of Europe. Except those two who let loose the whirlwind, the nations were all taken by surprise. England had made no preparations, but though unprepared she was not entirely unready: her fleet was armed and assembled, her expeditionary force was a real and not a paper force, and by Lord Haldane's brilliant reorganization her army system for the first time provided a framework capable of immediate and indefinite expansion. Within ten days of mobilization the expeditionary force was on the sea; within twenty it was in action on the field of battle; within a year it had been reinforced by ten times its original number of trained troops.

The 52nd left Aldershot on August 13 and embarked at Southampton that night; the transport sailed at 8 p.m., and reached Boulogne at 2 p.m. next day. The troops were received on landing with the utmost intensity of enthusiasm, and marched through admiring

and affectionate crowds to their quarters for the night, 1914
which were in the historic camp near Napoleon's monument on the heights above the town. They entrained for the front on the 16th, and spent a long day and night on the line, receiving at every station the fervent greetings of a people already threatened by the tide of conquest. They detrained at Wassigny in the early morning of the 17th, marched to Mennevret, and went into billets there. On the 21st they marched to La Groise, on the 22nd to Pont sur Sambre, and finally took up their assigned position at Genly, in the 5th brigade of the 2nd Division of General Sir Douglas Haig's 1st Army Corps. They were near the centre of the British line, in support of the troops which had arrived before them and were already entrenching between Mons and Binche. Smith-Dorrien's corps lay to their left, between Mons and Condé, the whole line being about twenty-five miles long.

The 52nd were stationed on a hill and had a bird's-eye view of the battle, in which they were not actively engaged. The German tactics were to shell the trenches first, then to lift on to the ground behind, in order to catch retiring troops or supports advancing, and to attack at the same time in massed formation. Their artillery was much helped by the smoke bombs dropped from aeroplanes on our most crowded points, but their infantry fire was quite ineffective and their charges unsuccessful. They came on in thick drifts, "like the crowd coming away from a football match," and were mown down in thousands without once reaching the trenches.

But the pressure became steadily greater even on the First Corps, for it had to bear the weight not only of Von Kluck's frontal attack but also of Von

1914 Buelow's advance on the right flank from Charleroi and the Sambre, where he had defeated the 5th French army the day before. Owing to confusion the news of this defeat and the consequent retirement only reached General French at 5 p.m., when the German attack was at its height. He perceived at once that with only 75,000 men, outflanked on both sides by more than twice their number, the position could not be held, and he began at once that retirement in conformity with the French Commander-in-Chief's movement which is now memorable as the Great Retreat.

The 52nd were at first moved up N.W. from Genly to Paturages. The roadway was cobbled, the day was sultry, and the march soon became a very painful trial of endurance. In the long hours of the night the men were gradually exhausted by the heat and the heavy burden of their packs. They kept doggedly on, with their feet bleeding and their backs bent nearly double, but exact order was impossible; the regiment struggled along, a solid shapeless mass of men, occupying the whole width of the road, their officers moving up and down the stream, trying to keep one platoon separate from another, cheering on their men, and refusing to give up any part of their own blistering equipment. By increasingly painful stages of two hours each Paturages was reached at last, and there, behind the town, the regiment entrenched itself, leaving two platoons to hold the streets in its front.

Within a few hours German shells began to fall on the town, and a fresh retirement was ordered. Behind Eugies two platoons of the regiment again dug themselves in, the heat becoming more and more

intolerable and the men more and more exhausted. 1914
At 7.30 a.m. the regiment moved again and narrowly escaped losing the two entrenched platoons. The order to retire did not reach them: they were left unsupported, with no other British troops in sight, and when the heads of the German columns became visible in the distance Captain Wood rode off to get what information or orders he could. He succeeded in finding the Brigade Major and learned that an orderly —who never arrived—had been sent some time before to say that the whole British Army was in retreat and already some miles ahead. Immediately afterwards, a mounted officer galloped into the rearmost platoons as they were forming in the road and warned them of the approach of a body of 500 Uhlans. The two platoons struck off to the west, and, to their great relief, found the main body of the regiment about two miles away.

The night march to Bavai was also one to be remembered. The regiment was now rearguard to the 5th Brigade, which was itself rearguard to the 1st Army, and B Company formed the rear point of the regiment. They marched with fixed bayonets, and halted and faced about at every cross-road through the night, the Germans being now on both their flanks. The road itself was an extraordinary sight, bordered by an endless moraine of abandoned greatcoats, equipment and entrenching tools.

At Bavai the regiment had two hours sleep; then marched in the early morning of the 25th past the Foret de Mormal to Pont-sur-Sambre, where they were told off to entrench and defend the four bridges of Aymeries, Aunoye, Berlamont and Sassegnies. These were handed over to the French next day, and

1914 blown up. At Berlamont the explosion took place before warning had reached Captain Godsal and 2nd Lieutenant Button; both of these officers were wounded and afterwards fell into the hands of the enemy, who captured the field hospital. The regiment moved south again from Barzy soon after midnight, and, with a few halts, marched till 6.30 p.m. on the 27th, reaching Neuvillette more exhausted than ever. They were only less unfortunate than Smith-Dorrien's men, who, after fighting their never-to-be-forgotten battle at Le Cateau till 3.30 that afternoon, were now making their escape from the pincers of Von Kluck by an even more painful and desperate march—perhaps the most critical ever made by a British army.

On the 27th the 52nd were relieved by the Glouces-ters, and the Brigade were all once more together. On the 28th they marched by La Fère to Servais, and there on the 29th they had their first day's rest—spent in washing clothes and sleeping while they dried. A French wagon was hired to carry the men's greatcoats.

The 6th French army had now got into position on our left, and the German advance seemed to be checked. But the pressure on the French centre was still heavy, and on the 30th the retreat began again. The 52nd reached Soissons that day and Laversine on the 31st. On September 1 the Guards (4th) Brigade beat off an attack at Villers-Cotterets; on the 3rd the British army reached the Marne, and on the 5th it was concentrated further south on the tributary called the Grand Morin. The Great Retreat was over and the British army was unbeaten: "the men are harder and more cheerful every day," wrote one of those who knew them best.

General Joffre's masterly order of September 4 1914 allotted four armies to the concentration on the extreme left, " to take advantage of the rash situation of the 1st German Army." Of these the British army was the second in order, and was to be posted " on the front Changis-Coulommiers, facing eastwards, ready to attack in the general direction of Montmirail." The offensive was to be taken by all on September 6, beginning in the morning. On September 6, therefore, Sir John French issued his order calling upon the British army " to show now to the enemy its power and to push on vigorously to the attack beside the 6th French Army." This was received by the men with the utmost satisfaction: not even Moore's army at Corunna were more pleased at being allowed to turn and attack their pursuers.

The battle which followed was a complete reversal of the fighting of the previous fortnight. The German right was outflanked and outfought by Manoury, their right centre was driven across four rivers in five days by French and Esperey, their centre was pierced in two places by Foch, and trapped in the marshes of St. Gond. The British drive reached Coulommiers on the 6th, passed Rebais on the 7th, Montceaux and La Trétoire on the 8th, La Ferté sous Jouarre on the 9th, and came in sight of the Ourcq on the 10th. The sharpest of the fighting was at La Trétoire on the 8th, where Sir Douglas Haig's corps repelled a desperate counter-attack and took a large number of guns and prisoners. After this the pace became faster: the German resistance was broken, and signs of demoralization began to appear. The British artillery drove on without waiting for support, and shelled the retreating masses almost without reply. To the 52nd, as they

1914 tramped joyfully up behind, the road presented much the same spectacle as that of their own retreat a fortnight back; but they saw it with very different feelings—the great-coats by the roadside were German great-coats, and there was proof that of this army many thousands would never turn to fight again.

The drive ended on the 11th, and on the 12th the battle of the Aisne began. The Germans were now in the immensely strong position which they had prepared in readiness for this precise emergency, and made a stubborn resistance along the line of the river. Sir John French ordered the crossing to be forced on the 13th, and his dispatch gives this account of it: "On the left the leading troops of the 2nd Division reached the river by 9 o'clock. The 5th Infantry Brigade [in which were the 52nd] were only enabled to cross in single file and under considerable shell fire, by means of the broken girder of the bridge, which was not entirely submerged in the river." The attack was only partially successful, and "the 2nd Division bivouacked as a whole on the southern bank of the river, leaving only the 5th Brigade on the north bank to establish a bridgehead." On the 12th the 52nd captured 7 officers and 107 men, some of them Landwehr, some Uhlans of the Guard.

On the 14th the 52nd established themselves in the village of Moussy, and prepared for the attack on the advanced part of the German position. The enemy were holding the crests and slopes of the hills which jut out towards the river in a series of spurs with valleys intersecting them. A frontal attack on the spurs was almost hopeless; the 52nd made their first attempt by a lateral movement, advancing up the thickly wooded slope of one of the intermediate valleys, and digging in on the left-hand side of it, at right angles to the river. But here they were soon enfiladed by fire from a trench across the head of the valley, and eventually they were withdrawn. Meanwhile the Connaught Rangers had twice charged up the next headland, above Soupir, and twice been repulsed by counter-attacks, after doing great execution. The Coldstream Guards were then thrown in, and with the gallant remnant of the Connaughts, who refused to be left behind, they stormed the position and dug in on the crest of the hill. 1914

During this attack a battery of Field Artillery at Moussy was ordered to take up a position some way to the right. The only road was that which ran along the river side and crossed the mouth of the valley where the 52nd had been enfiladed. The Germans had now got their guns into action and were bursting shells on the exposed part of this road; but the artillery officer, by sending his guns one at a time and not sparing the horses, got his whole battery through with comparatively few casualties. This was a most inspiriting sight and a timely hint to the 52nd. An urgent message came down at this moment from the Coldstream Guards, who were hard pressed by a counter-attack on the crest of the Soupir hill. Between Moussy and Soupir lay the exposed stretch of road. Colonel Davies had watched the gunners pass it and saw that there was a chance even for infantry, though not on the road itself. He formed up the regiment under cover of the Moussy spur, and sent them in single file and at top speed along the strip of grass between the road and the river. The shells continued to burst with perfect accuracy upon the road, but the runners were beyond and below the road, and got across with wonderfully little loss. Their casualties on this day

1914 were 4 men killed, and Lieutenant Owen and 40 men wounded, all by shell fire. They climbed the Soupir ridge to find the counter-attack dying away, and the trenches which they then occupied along the Cour de Soupir were the home of the regiment for weeks to come. "Throughout the battle of the Aisne," wrote Sir John French, "this advanced and commanding position was maintained, and I cannot speak too highly of the valuable services rendered by Sir Douglas Haig and the Army Corps under his command. Day after day and night after night the enemy's infantry has been hurled against him in violent counter-attack, which has never on any occasion succeeded, whilst the trenches all over his position have been under continuous heavy artillery fire."

From this heavy fire the 52nd suffered one very severe disaster. On September 16, near the Cour de Soupir, a single big shell killed Lieutenants Worthington and Mockler-Ferryman, 2nd Lieut. Girardot and 28 men, and wounded Captains Higgins and Evelegh, 2nd Lieut. Tylden-Pattenson and 8 men. On the 19th Captain Evelegh was again hit and killed, with 8 men, and Captain Blewitt, 2nd Lieut. Barrington-Kennett, and 24 men were wounded.

At the beginning of October it was decided to move the British army from the Aisne and send it north to support the extreme left flank of the Allies, which was now seriously threatened. The transference of the trenches to the French was begun on the night of October 2, and on the 3rd Smith-Dorrien's corps moved off. The operation was a delicate and dangerous one, and the utmost secrecy was observed. The 1st corps was the last to be withdrawn; when it detrained at St. Omer on the 19th the others had already been fighting hard for a week to hold a very 1914 extended front. Sir John French had now to make a critical decision—whether to use the 1st corps as a reinforcement for the hard-pressed line already formed, or to leave Smith-Dorrien to hold on as best he could and send Haig's corps into the yawning gap between the British left and the sea, where only the remnants of the Belgian army and a few French Territorials were heroically awaiting the German attack. He states the problem and his solution of it in stirring words:

"After the hard fighting it had undergone, the Belgian army was in no condition to withstand, unsupported, such an attack; and unless some substantial resistance could be offered to this threatened turning movement, the Allied flank must be turned and the Channel ports laid bare to the enemy. I judged that a successful movement of this kind would be fraught with such disastrous consequences that the risk of operating on so extended a front must be undertaken; and I directed Sir Douglas Haig to move with the 1st corps north of Ypres. . . . No more arduous task has ever been assigned to British soldiers, and in all their splendid history there is no instance of their having answered so magnificently to the desperate calls which of necessity were made upon them."

The 1st corps aimed at a line between Poelcapelle and Paschendaele, but were held up on the line Zonnebeke—St. Julien—Langemarck—Bixschoote. They reached this position on October 21, and were ordered to hold it at all costs until relieved—probably on the 24th—by the French. In the fighting of the 21st the 52nd were in action near Langemarck. The regiment marched at dawn, and at 9 a.m. was ordered to attack a strongly entrenched position, in front of which lay a

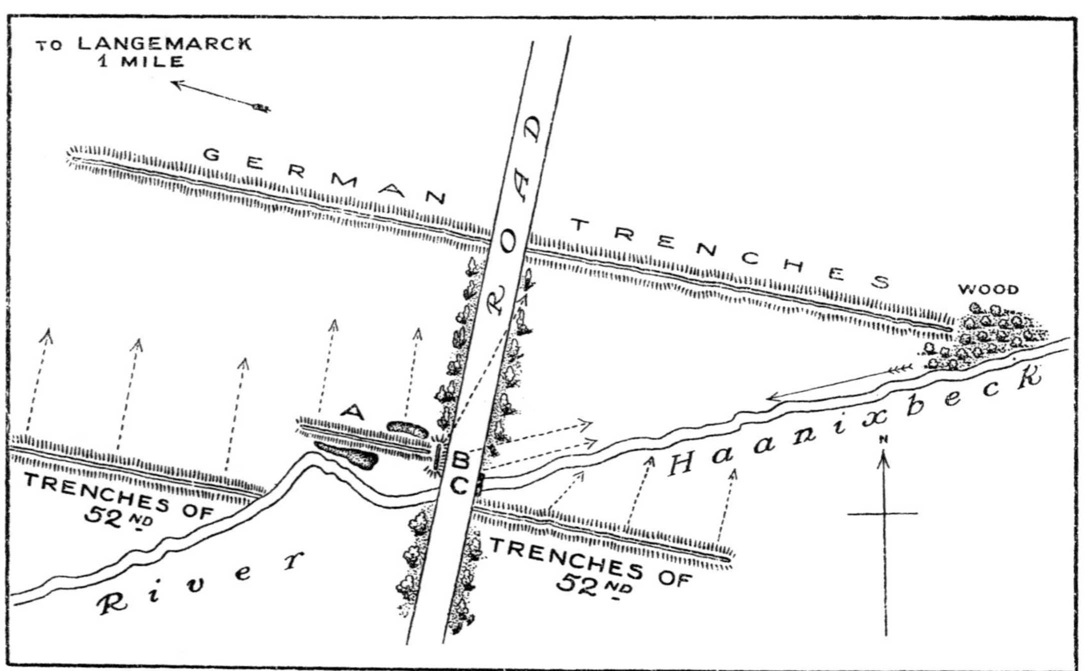

POSITION OF THE 52ND, OCTOBER 21–24, 1914.

A, *Advanced trench with shell craters* ; B, *small trench to command snipers and river bank* ; C, *culvert*. ***The dotted arrows show the field of fire.***

high thick hedge fortified with barbed wire and swept 1914
by machine guns. In this hedge a single wide gate seemed to offer the only chance of passing through; but it was accurately ranged by the Germans, and of the many officers who tried it every one was mown down except Captain Tolson and Lieut. Spencer. By the most desperate bravery the hedge was then broken—ten[1] officers and 200 men had fallen in a quarter of an hour—and the remainder of the regiment raced across the open bullet-swept slope towards the German position. The men's blood was up and they were bent on reaching the German trenches with the bayonet. But away on the left the 1st Division had been held up far to the rear, and the 52nd, if they had advanced further, must have been enfiladed or even taken in reverse. The order was therefore given to dig in where they were, the *point d'appui* thus formed being connected with the general line of the 1st Division by the sloped-back trenches of the Highland Light Infantry.

They had now to hold on for three days and nights, and the skill and tenacity by which they accomplished this can only be realized if the position is understood in detail. The German trenches ran at right angles across a straight road bordered with tall poplars. The trenches of the 52nd were also on both sides of this road, but they were prevented from running continuously by the course of a small stream—the Haanixbeck—which flowed from behind our left, crossed the road by a culvert, and went on to join the German

[1] Killed, Captain Harden, Lieutenants Murphy and Turbutt, 2nd Lieutenant Filleul; wounded, Major Eden, Captains Wood and Kirkpatrick, 2nd Lieutenants Marshall, Newton-King, and Chippendale. 2nd Lieutenant Marshall died two days later.

1914 extreme left at an angle of thirty degrees. The companies of the 52nd were no longer separate, and officers took command of whatever men happened to be around them. Away to the left, with their backs to the stream, Captain Ponsonby and his men dug in. Twenty yards in advance and still on the left of the road was a forward trench with Captain Kirkpatrick and Lieutenant Baines in command; while on the right of the road, but some twelve yards in rear, and behind the stream, 100 men under Captain Dillon were entrenched. The small forward trench was of importance as covering the gap between the other two, but it was in a peculiarly difficult position, for besides being fully exposed to the enemy's front, it was peppered at closer range by snipers behind the poplars on the road, and enfiladed by fire from a small wood which filled the angle where the stream met the German left down the valley. Moreover an attack from the wood in this angle would have a clear run up the valley against the flank of the trench.

These dangers were pressing and were dealt with at once. In spite of an incessant bombardment by the enemy's howitzers, an advanced pit was dug for two sharpshooters, who fired down the road and had great success in picking off the snipers. To command the valley and wood a trench for ten men was dug at right angles to the position and close to the road and stream. This was supplemented, on his own initiative, by Private Hastings of B Company, who climbed down into the culvert by which the stream passed under the road and lay there for three nights and two days, shooting incessantly down the valley as the Germans tried incessantly to creep up it. From this position he emerged only once; when it became evident that the German wounded in front of him could expect no help from their own side, he fetched a wheelbarrow from the road near the enemy's trenches and brought them in himself across the open. Even without the Distinguished Conduct Medal which he received, the regiment will have no difficulty in remembering Private Hastings.[1] 1914

The Germans attacked constantly and furiously both by day and night. They came on in big rushes, and the mixed companies of the 52nd had to keep up a very rapid fire. But it was effective—after the first night attack 70 dead Germans were lying within fifty yards of this part of the line—a front of only 100 yards. On the second day the German howitzers with their "Black Marias" systematically destroyed first the right and then the left hand of Lieutenant Baines's forward trench, burying him to the waist and killing many of his men. He withdrew the survivors to the river bed until dusk, and then at once converted the shell craters into fresh trenches. The enemy lost no time in attacking, and the 52nd were firing into the surging masses of them all night. During this attack the 52nd had two officers—Captain Ponsonby and 2nd Lieut. Humfrey—wounded, and two men killed and five wounded: the Germans left 670 dead on the ground. Towards morning the attack died away, and when the French came up, true to their day, they found the position still intact. Sir John French had been obeyed "at all costs," and another splendid page had been added

[1] Nor will they forget Bugler Lovelace, who in the Battle of the Aisne volunteered to fetch water under a heavy fire; and Private Kippax, who won the *Médaille Militaire*, by attacking a German patrol single-handed, shooting the officer and six men, and turning their machine gun to cover the advance of the 52nd.

1914 to the record of the 52nd. In the memory of the survivors certain gallant figures stand out with unfading clearness—Ponsonby bringing up reserves in a storm of bullets, Kirkpatrick holding on to his forward trench after being twice wounded in the neck and side, Baines coolly cutting the barbed wire in sight of the German rifles, Southey with his machine guns steadily rattling the enemy, and " the men "—always, their officers said, " absolutely magnificent."

But relief did not mean rest; if Calais was to be saved every man was needed in the line. The 52nd, when they had handed over to the French on the 24th, were sent immediately to the Zandvoorde-Gheluvelt line to relieve the 7th Division, which had crossed Belgium alone in face of several German army corps and was now holding a very extended front with terribly reduced numbers.

That day the Germans drove in the point of the British salient and entered the Polygon wood. A counter-attack on the 25th took some guns and prisoners but failed to clear the wood. At night a fresh attack by the Germans broke through the salient at Kruseik; the 2nd Scots Guards sacrificed themselves in vain, but the 7th Cavalry Brigade saved the situation. On the 26th and 27th our line was readjusted: the remnant of the 7th Division was ordered to be temporarily incorporated with the 1st Division, of which on the 30th the 52nd also became part for the time.

The Kaiser had now reached the front with enormous reinforcements and ordered the capture of Ypres, which, as he told the Bavarians, would determine the issue of the war. At 5.30 a.m. on the 29th an intercepted wireless message gave warning of the attack, and an hour later it was launched at the point where the 1st and 7th Divisions were holding the Gheluvelt cross-roads. The 1st Division was pushed back for a time, but in the afternoon the enemy gave way and at dusk the Kruseik ridge was ours again. At dawn on the 30th a fresh attack was made towards Klein Zillebeke, held by Byng's Cavalry, and the Zandvoorde ridge, held by the 7th Division. Success at this point would have meant the isolation and destruction of the whole of the 1st corps. Sir Douglas Haig ordered that the line should be held at all costs, and borrowed reinforcements from the 2nd corps and from the 9th French army. 1914

On the 31st came the climax—" the most important and decisive attack," says Sir John French, " made against the 1st corps during the whole of its arduous experiences in the neighbourhood of Ypres." The dispatch needs several corrections or alterations of detail, for in the description of the battle no account is taken of the position of the 52nd, two companies of which (A and B, Captain Dillon's and Lieutenant Baines's [1]) were posted between the right of the 22nd Brigade and the left of the 2nd Brigade east of Klein Zillebeke. When, therefore, the 22nd Brigade were driven from their trenches by the tremendous weight of the German attack, it was not the 2nd Brigade, but the two companies of the 52nd, who found their left flank exposed. Fortunately the Germans were too intent on occupying the vacant trenches to take any notice of the 52nd, who held stoutly on, first firing over their left shoulders at the crowd that was passing them, and then lining a hedge at right angles to the end of their trench.

[1] The other two companies were digging trenches on Hill 60, a mile to the rear.

1914 Further, the left of the 2nd Brigade was covered not only by the 52nd but by the Irish Guards, who were also ignored in the dispatch, no doubt because only "unofficially" present as emergency reinforcements. They too were shortly afterwards compelled to give ground (though not from any exposure of their flank), and the two companies of the 52nd, to their consternation, found themselves with both flanks in the air.

They were then withdrawn and ordered to conform to the rest of the line. They fell back behind a wood into which the Germans had poured over the trenches of the 22nd. There they joined with the Northamptonshire Regiment, and by a sweeping charge succeeded in clearing the enemy out of the wood and establishing a sloping line up to their own old trenches, which they thus re-occupied. They were probably, therefore, the troops mentioned in the dispatch as "the right of the 7th Division, who managed to hold on to its old trenches till nightfall."

Next day, November 1, the 52nd had their whole force in the line—it now only amounted to 380 men out of the 1100 who had entrained from Aldershot. B Company, numbering 78, was in close support; the other three held some of the new trenches on the left, their own old trenches in the centre, and half the old trenches of the Irish Guards on the right. The remnant of the Irish Guards were again decimated and driven back by a heavy bombardment, and the right flank of the 52nd—C Company—was completely exposed. The commander of the company, Lieutenant Tylden-Pattenson, was equal to the occasion. He was a very young officer, who had only joined three months before; remarkable, like Wellington's favourite young men, for the smartness of his equipment and the cheerfulness of his spirits. He had done admirably in keeping his men up during the Great Retreat, and now in the moment of extreme danger both he and they appeared perfectly happy, firing away without a thought of giving ground. When the order to retire came, Lieutenant Tylden-Pattenson sent his men back one by one through the storm of shrapnel, but many fell before the wood was reached. He himself was not hit—he had yet some months of service to give his country. 1914

On November 2 the regiment was removed from the 7th Division and attached to the 4th (Guards) Brigade. The line was wearing very thin, and there were no reserves: patching was all that could be done, and the patches themselves were only rags—but rags of the finest stuff ever woven. The 52nd that day had been digging second-line trenches, which were urgently needed. Determined to finish before resting, they were digging on into the night, when an order came for them to go forward and relieve the Grenadiers. They found them in small and inadequate trenches, and completely exhausted. By working all through the hours of darkness the 52nd greatly improved the trenches and even got some barbed wire up: they also brought in a number of wounded Germans and some other booty—one company accumulated altogether 33 German rifles and 2500 rounds of ammunition. One officer with two men crept out in front to examine some mysterious little clouds of smoke which looked like bombs; they found to their astonishment that the smoke came from the packs of dead Germans, the contents of which were, for some reason, slowly smouldering.

1914 During the next two days and nights the 52nd worked hard to make their trenches in every way perfect, with traverses, communication-trench, wire-entanglements and various devices against surprise, including arrangements of beef-tins and jam-pots, and a night post of fir-poles and straw across a broken road, where two men could lie and fire. On November 3 some Germans began digging in a hollow only thirty yards in front of A Company's trench. 2nd Lieuts. Pepys and Pendavis, with Privates Hall and Merry, stood up on the parapet and fired on them with machine guns, and finally advanced and drove off the survivors. This was cheering, and the men took a pride in their work, which helped to sustain them, but it became more and more difficult to speak of reinforcements with the inward certainty that there were none to come. The officers were falling one by one; on the 3rd Lieutenant Rupert Brett was wounded by shrapnel in the head, on the 6th fell Lieutenant Ward, who had only joined two days before. He was with part of B Company in support while the rest of the regiment was hotly engaged in front. An urgent request came from another regiment for volunteers to take a hedge which the Germans had lined. Lieutenant Ward threw away all his equipment but his revolver, led his men magnificently to the charge, and there ended his "one crowded hour of glorious life." His company commander, Lieutenant Baines, when he heard of what had happened, ran out alone and found him dead by the hedge: he tried to bring him back, but that was not possible till night. The names of both officers were mentioned in the same dispatch.

The climax of this stupendous battle came, after three days of comparative quiet, on Wednesday,

November 11. For the final stroke at Ypres the Kaiser 1914 had brought up the 1st and 4th Brigades of the Prussian Guards—thirteen battalions in all, including the 1st and 3rd Foot Guards, the Kaiser Franz Grenadier Regiment No. 2, and the Koenigin Augusta Grenadier Regiment No. 4. In the twilight of early morning the huge column advanced with all the pomp of their parade step against our salient at Gheluvelt.

At this moment the 52nd, who had been relieved and sent north on the 9th, were in reserve at Verloerenhoek, on the Ypres-Zonnebeke road. On the morning of the 11th they were unpacking their equipment for the first time for weeks, and preparing for a rest, when an urgent message reached them: "The line is broken—the Prussian Guards are through." The few available supports were desperately needed. The 52nd covered the two miles to Externest as fast as they could go: they arrived to find a strange and bewildering scene before them. In the angle of the cross-roads near the rear of the Nonneboschen wood they saw some French guns, silent and apparently deserted; across the road behind them were some English guns, also deserted—their gunners were deployed in front with rifles—"the only men," said their commander, "between the Germans and Ypres—thank God you've come!"

The wood itself was full of Prussians, who had broken the 1st Division by sheer weight and flowed over their trenches; many of them were now visible on the near edge of the wood, but they seemed uncertain of their direction and they had not yet discovered the silent guns. Colonel Davies had the chance of a hundred years before him, and he took it on the instant. He established the Regimental Headquarters on the

1914 right front of the French guns, at the point where the road ran nearest to the rear angle of the wood; D Company (Captain Tolson and Lieutenant Vere Spencer) he stationed for the moment behind the guns as a reserve. B Company (Lieutenant Baines) and C Company (2nd Lieutenants Tylden-Pattenson and Titherington) were to charge, while A Company (Captain Dillon, Lieutenant Pepys and 2nd Lieutenant Pendavis) kept up a covering fire on their left flank.

The four companies mustered perhaps 350 in all; of the Prussian Guards there were about 800, and the officers of the 52nd as they charged were struck by the immense superiority of the enemy in physical bulk—our men appeared to be only half their size.[1] But there the superiority ended: the Prussians had already met the 1st Guards Brigade, and though their weight had carried them through the trenches they had lost their sense of direction, their cohesion, and some part of their resolution. They were now face to face with the finest Light Infantry in the world: it is little shame to them that their courage and their discipline were not equal to their need. In the hundredth year since Waterloo the 52nd were not out to flinch or fumble; they manœuvred and fought with the swift precision which alone could honour the memory of Moore and Colborne.

The open ground to be covered was some 300 yards; as Baines and Tylden-Pattenson crossed it with their slender converging lines the enemy had their chance, but Dillon's fire pinned them in their covert; then when the two companies had rushed the edge of the wood and were entering the dense undergrowth, he joined in

[1] Lieut. Titherington buried three who were over seven feet in height.

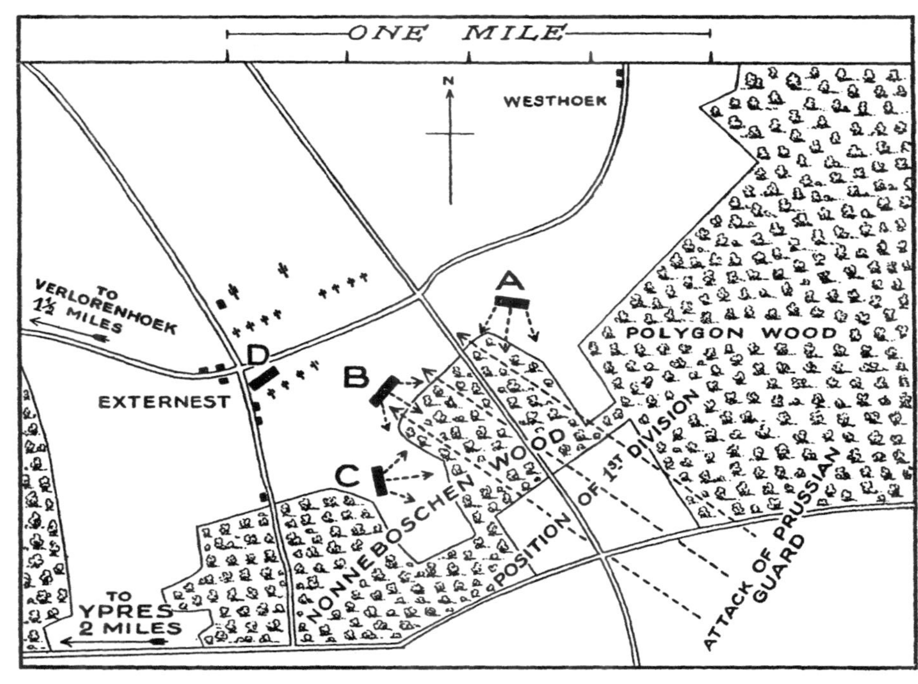

CHARGE OF THE 52ND, NOVEMBER 11, 1914.

A, B *and* C, *Companies of the 52nd attacking the Prussian Guard*; D, *Company at first in reserve behind French guns.*

1914 on the left, and as the thin line went forward, stretched to its utmost, Tolson came on with his company as second line. The whole attack went with the old Light Division click; even the wood of Redinha was not cleared " in more gallant style " than this. The giants made no effective stand; the drive was carried through without a check, our men enjoying it, said one of them, "as if we were all beating the wood for pheasants, at the double." The comparison was curiously apt: for an officer of the 1st Division has described how a remnant of the 1st Brigade, hearing the firing in the wood, had posted themselves by their old trenches on the other side, and waited, like the line of guns at a " hot corner." When the beaters approached, the first sign of life was a rise of pheasants on the edge of the wood, followed by a rush of a few Germans, who were all shot down as they left the covert. The 52nd saw nothing of this; they drove straight out to the front, and found there a great number of killed and wounded, with a few scattered men of the 1st Division, and further off a confused mass of Prussians occupying trenches under cover of artillery fire. Here, when the wood was practically cleared, B Company lost their only officer, Lieutenant Baines, wounded by shrapnel in the right shoulder. He was able, however, to walk the two miles back to Regimental Headquarters, escorted by a Prussian officer and five other prisoners of the Guards, who carried his equipment for him. To an English officer of another regiment, who met them at the cross-roads, this procession was, perhaps, the most surprising sight of his life.

The front companies of the 52nd were now joined by some of the Northamptonshire regiment on the

right and some of the Connaught Rangers and the 1914
5th Field Company, R.E. on the left. Led by Captain Dillon, they charged the Germans, and took one line of trenches, with some prisoners. They would have taken the second line too, but for the fire of the French guns, which kept shelling the trenches until dark, in ignorance of our progress. The regiment was now collected and entrenched for the night, to the west of the Polygon Wood. The casualties for the day were extraordinarily slight: 2nd Lieut. Jones and four men were killed, and Lieutenant Baines and seventeen men wounded.

With this counter-attack of the 52nd the crisis of the first battle of Ypres had passed; the Kaiser's final attempt of November 17 never came so near to success. The fight of the 11th assured us of victory, and the victory was in the main our own. It was made possible by the co-operation of the French and Belgians, "and no allies," says Mr. Buchan, "ever fought in more splendid accord. But the most critical task fell to the British troops, and not the least of the gain was the assurance it gave of their quality. They opposed the blood and iron of the Germans with a stronger blood and a finer iron. . . . The steady old regiments of the line revealed their ancient endurance." That is well said: it is enough. A regiment like the 52nd cannot surpass their ancient record; but they can add to it, and keep it fresh in the memory and admiration of their countrymen. When the first lists of honours appeared in 1914, those who know the history of the Light Division noted with a familiar pride that of the eight company officers who led the 52nd at Ypres five received the Distinguished Service Order and another the Military Cross. No other single battalion

1914 in the army equalled or nearly equalled this achievement; yet even these are but the honours of the living, and they are no brighter than the unclaimed honours of the dead.

Nor, perhaps, is it any honours that the officers of such a regiment most desire. It is easy to believe that they would value most the acknowledgment, in a few plain words from a tried commander, that their great tradition had been kept. If so, the 52nd have already had their wish. When Major-General Haking, C.B., relinquished the command of the 5th Brigade, he wrote a letter of thanks to the regiment in these words:

"The rapid and skilful manœuvring of the battalion during the retirement from Mons and the subsequent advance to the Aisne, their defence during the long occupation of the latter, and, above all, their splendid attacks and defence round Ypres, are well known throughout the whole army, and will later on become a matter of history. The battalion has always been celebrated for its attack at Waterloo, but in my opinion it will in future be distinguished above others for its magnificent attack near Ypres. . . . I cannot tell you what satisfaction it gives me to be able to record in this brief manner the heroic doings of the battalion during the present campaign, the value of which cannot be exaggerated."

INTRODUCTION TO THE APPENDICES

It would evidently be impossible, during a war like the present, world-wide and long continuing, to write even the story of a single regiment with any accuracy, unless the narrative were confined to a period already closed. It was thought best on this occasion to deal only with the Western Campaign during 1914, and in the case of the Oxfordshire and Buckinghamshire Light Infantry the result is to limit the fighting record to the 2nd Battalion only. It is hoped that a future opportunity may be found of continuing the story through 1915, and thereby including an account of the successes of the 1st Battalion in the brilliant campaign in Mesopotamia, and of the Territorial and Service Battalions elsewhere. For the period covered by the present volume the regiment is represented by the 52nd.

We may, however, get from the two Appendices which follow an idea of the wider service rendered by the regiment as a whole, even in the earlier part of the war. Under the present system a regiment is a much more powerful and important part of the national forces than it could be in the days when it was an isolated unit, one or two battalions strong, with no further means of expansion. The first of our Appendices will show this at a glance. From the headings there given it will be seen that the 43rd and 52nd, though now as always the main strength of the Oxfordshire and Buckinghamshire men and their historic pride, are actually the first line of an Imperial force numbering in all ten battalions in the British Army, and two allied regiments in Canada and New Zealand. The state of the regiment immediately before the war was as follows. After the 1st and 2nd Battalions (Regular) came the 3rd Battalion, representing the old Oxford Militia, and now called the Special Reserve: then two Territorial Battalions, the 4th and the Buckinghamshire, with headquarters at Oxford and Aylesbury. The allied regiments were the 52nd Regiment of Canadian Militia (Prince Albert Volunteers) with headquarters at Prince Albert, Saskatchewan, and the 6th (Hauraki) Regiment of the Dominion of New Zealand.

Upon the outbreak of war the expansion provided for by the Haldane system took immediate effect. First, the establishment of the existing battalions was largely increased by the commissioning of second-lieutenants from the O.T.C. and from Oxford University. Of these thirteen had joined the 3rd Battalion within a few days of mobilisation, and before September 1915 the number serving had increased to sixty-six. In the same time sixty-eight were added to the 4th Battalion and sixty-one to the Buckinghamshire Battalion. A considerable number of these have already seen active service—the 3rd Battalion has sent more than twenty second-lieutenants to the front, to the 2nd Battalion or to other regiments in need of assistance. As a further expansion there were added to the regiment upon the raising of the New Army four Service Battalions and a 2nd Reserve Battalion. Of these the 5th, 6th, 7th, and 8th Battalions have an establishment on the ordinary scale, while the Reserve has on its list, besides seven captains and eleven lieutenants, some of whom have already gone to the front, no less than 114 second-lieutenants on probation and fifteen more "attached." Lastly, it must be remembered that the recruiting of the regiment has not only furnished the four new Service Battalions, but has enabled the Special Reserve and 2nd Reserve to make good, by drafts, the losses in the field, which, in the case of the 2nd Battalion, have been large beyond all expectation.

Appendix A, then, shows the state of the regiment after the expansion has taken place; but it is in one respect incomplete. Its first page lacks no less than twenty-seven names, the names which a year of war has transferred from the active list to the historic strength of the regiment. These will be found in the first column of Appendix B—the Roll of Honour. Nineteen of them are the names of officers killed while serving in the 52nd—a very heavy list for a single battalion, and in marked disproportion to the number of wounded, which may be put at thirty-five for the same battalion. This gives a total of fifty-four officer casualties for the 52nd alone, but even this is not all, for at least four officers have been wounded on two separate occasions, and of those killed no less than five had returned to the front after being wounded a first time. The real total is, therefore, sixty-three casualties, or about double the number of the officers with the battalion when it first landed in France.

This is by no means an exceptional figure in the present war—many regiments have suffered as severely, and some even more severely—but it is none the less remarkable. The continued efficiency of the battalion after a year of such losses is only to be accounted for by the power of the regimental tradition, and the skill and devotion of the officers responsible for the training of the regiment, whether at the front or in the barracks of the 3rd Battalion.

Of the 1st Battalion Appendix B includes only one officer killed and two wounded; but though their losses in the first part of the war may have been small, the achievement of the battalion, when the story of the Mesopotamian campaign can be fully told, will be found to be singularly consistent with the historic record of the 43rd. The other battalions of the regiment will have their honour too, when all is known: at present it is only possible to say that the 4th Battalion (T.F.) has lost three officers killed and three wounded, the Buckinghamshire Battalion (T.F.) five wounded, and the 5th (Service) Battalion two killed and nine wounded. There are also among the casualties the names of eight officers of the regiment who have been attached to other corps less well supplied with reserves of their own.

When the time comes, a third Appendix might well give the names of those officers and men of the regiment who have been mentioned, promoted or decorated for their services. At present such a list could be made out for the 52nd only: but it would be a remarkable one, for it would enumerate more than thirty marks of distinction for the officers of a single battalion. Lieut.-Colonel H. R. Davies, who commanded in every action here recorded, with a fine courage and judgment worthy of his famous predecessors, has been twice mentioned in Dispatches and twice promoted. When he left the battalion in March 1915, on becoming a Brigadier-General, he was succeeded by another distinguished officer, Major A. J. F. Eden,[1] who has also been mentioned in Dispatches and promoted to Brevet Lieut.-Colonel. Brevet majorities have been given to three Captains—J. T. Weatherby, E. H. Kirkpatrick, and H. R. M. Brooke-Popham (*Flying Corps*). The D.S.O. has been conferred on Captain and Adjutant R. B. Crosse, Captains H. M. Dillon, J. T. Weatherby, F. V. Holt (*Flying Corps*) and H. R. M. Brooke-Popham (*Flying Corps*), Lieutenants A. V. Spencer and C. S. Baines, and Second-Lieutenants F. Pepys and H. V. Pendavis. Captains G. Blewitt and W. G. Tolson and Lieut. E. H. Whitfeld have received the Military Cross, and Lieut. R. G. Worthington the Legion of Honour. Most of these officers have also been mentioned in Dispatches; and so too have Second-Lieutenant J. P. M. Ward; and Captain V. J. D. Bourke (*Flying Corps*), and the following privates and non-commissioned officers:—Privates Payne, Kippax, Hart, Salmon and Wheeler; Lance-Corporals Apsey, Dobson, Kerswell, Radley and Atkins; Sergeants Fossey, Waters and Wood.

The following have received the Distinguished Conduct Medal:—Privates Hall, Hastings, Merry, Stock, Tyrrell, Jones, Wykes, Zeacle; Corporal Hodges; Sergeants Ashby, Edwards, Breach and Colour-Sergeant Hudson. The French Médaille Militaire has been conferred on Corporal Kippax; the Russian

[1] Major H. L. Ruck-Keene, D.S.O., senior Major of the Regiment, was invalided at the time.

Medal of St. George on Privates Hastings and Tyrrell, and the Cross of St. George on Sergeant Dobson.

Brilliant as it is, this list still fails to do the 52nd justice, for in more than one instance death has forestalled the inevitable honour. Among those most distinguished by the admiration of their comrades, the regiment will remember Captain Ashley William Neville Ponsonby, a soldier who added to perfect courage an invaluable coolness and clearness of perception.

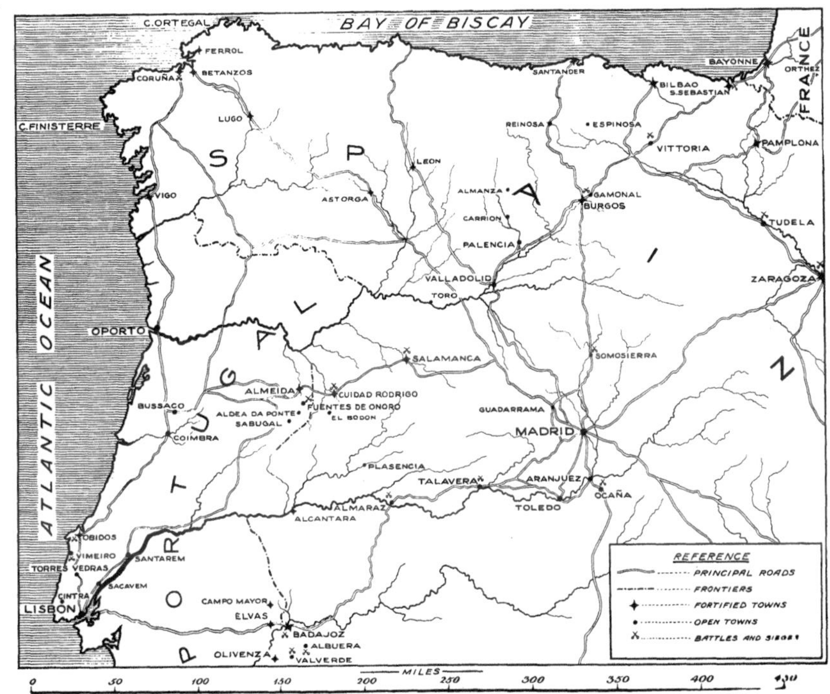

THE PENINSULA

Appendix A

LIST OF OFFICERS

SERVING IN

THE OXFORDSHIRE AND BUCKINGHAMSHIRE LIGHT INFANTRY

AT THE BEGINNING OF THE SECOND

YEAR OF THE GREAT WAR

This list is reprinted with the sanction of the War Office from the Army List of September 1915 *(corrected to August* 31*st,* 1915). *By kind permission of the Controller of His Majesty's Stationery Office the official stereos have been used.*

THE OXFORDSHIRE AND BUCKINGHAMSHIRE LIGHT INFANTRY.

Regimental District, No. 43. [No. 7 District.]

The United Red and White Rose.

"Quebec, 1759," "Martinique, 1762," "Havannah," "Mysore," "Hindoostan," "Martinique, 1794," "Vimiera," "Corunna," "Busaco," "Fuentes d'Onor," "Ciudad Rodrigo," "Badajoz," "Salamanca," "Vittoria," "Pyrenees," "Nivelle," "Nive," "Orthes," "Toulouse," "Peninsula," "Waterloo," "South Africa, 1851-2-3," "Delhi, 1857," "New Zealand," "Relief of Kimberley," "Paardeberg," "South Africa, 1900-02."

Agents—Messrs. Cox & Co.

Regular and Special Reserve Battalions.

Uniform—Scarlet. *Facings*—White.

1st Bn. (43rd Foot)
2nd „ (52nd Foot) | 3rd Bn. (Oxford Mil.) | Depôt .. .. .. *Oxford.* Record Office .. *Warwick.*

Territorial Force Battalions.

4th Bn. .. .. *Oxford.* | Buckinghamshire Bn. .. *Aylesbury.*

Service Battalions

... 5th Bn | 6th Bn. | 7th Bn. | 8th Bn.

2nd Reserve Battalion

.. .. .. 9th Bn.

Allied Regiment of Canadian Militia.

52nd Regiment (Prince Albert Volunteers.) *Prince Albert, Saskatchewan.*

Allied Regiment of Dominion of New Zealand.

6th (Hauraki) Regiment.

Colonel .. .. .. Colvile, Hon. Lt.-Gen. *Sir* F. M., *K.C.B.*, ret. pay [R] 14Sept.13

Officer Commanding Depôt Cuyler, Lt.-Col. (*temp.*) *Sir* C., *Bt.*, Maj. Res. of Off. .. 6Aug.14

Maul, Maj. H. C., T.F. Res. .. .. .. 13Mar.15 | Knox, Capt. (*temp.*) H. V., *late* Capt. R. Garr. R. 8Jan.15 8Sept.14

Adjutant .. .. Pearson, Maj. C. C., Res. of Off. 6Aug.14

Quarter-Master .. Carnell, Qr.-Mr. (*hon. capt.*) C., Res. of Off. 8Aug.14

Attached .. .. Maul, Capt. H. C., 3 Bn. .. .. .. —

1st and 2nd Battalions. (Regular.)

Lt.-Colonels.

s. 2Davies, H. R. (*temp. Brig.-Gen.*) 18Sept.11 *bt.col.* 18Feb.15
1Lethbridge, E. A. E., *D.S.O.* 25Oct.13

Majors.

2Ruck-Keene, H. L., *D.S.O.* 14Sept.07
(3) 2Darell-Brown, H. F. 18Sept.07
5Cobb, C.H. 18Sept.07 29Nov.00
1Hyde, A. C. 16Dec.08
2Eden, A. J. F. 18Sept.11 *bt. lt.-col.* 3June15 (*temp. lt.-col.* 26 *Mar.* 15)
s. Marriott Dodington, W., *p.s.c.* 23Oct.11
1Carter, L. J. 23June13
s. Pope-Hennessy, L. H. R., *D.S.O.*, *p.s.c.* [*l*] 25Oct.13 26Mar.06

Captains.

1Stapleton, F. H., *p.s.c.* 11Sept.02
Clayton, E. R., *p.s.c.* 11Sept.02
r *Frith, C. H.* 22Nov.02
1Forrest, C. E., *D.S.O.* 22Feb.03
Ballard, J. A. 23June03
1Henley, C. F. 8Oct.04
Bayley, A. G., *p.s.c.* 8Oct.04
s. *Brooke-Popham, H. R. M.*, D.S.O., *p.s.c.* [F] 9Nov.04 *bt. maj.* 3June13
s. *Wood, H. L.* 14Apr.05
s. *Fuller, J. F. C.* 21June05
w.a. *Keyworth, R.D.* 9Oct.05
Weatherby, J. T., D.S.O., *p.s.c.* 30Dec.05 *bt.-maj.* 18Feb.15
s. *Sullivan, G. A.* 28Dec.08

Captains—contd.

1Simpson, R. V. 28Dec.08
2Kirkpatrick, E. H. 13Feb.09 *bt.-maj.* 18Feb.15
2Higgins, C. G. [*l*] 13Feb.09
e.a. *Tunnard, L.* 22Jan.10
e.a. *Pollock, J. A.* 22Jan.10
2Ponsonby, A. W. N. 22Jan.10
2Dillon, H. M., *D.S.O.* 17May10
(5) 1Logan, R. O. 19Nov.10
1Morland, W. E. T. 22Jan.14
c.o. *Parr, B. C.* 14Mar.14
Stephens, R. 14Mar.14
1*Foljambe*, Hon. *J. C. W. S.*, *Adjt.* 14Mar.14
t. *Bartlett, A. J. N.* 22Mar.14
f.c. *Bourke, U. J. D.* 18Apr.14
s. *Blewitt, G.* 18Apr.14
f.c. *Holt, F. V.*, *D.S.O.* 25July14
(5) Sanderson, A. E. 25July14
(5) Carfrae, C. F. K. 5Aug.14
(2) Scott, L. F., *Res. of Off.* —
(2) Hammick, S. F., *Res. of Off.* —
(2) Chichester, C. O., *Res. of Off.* 5Aug.14
(2) North, A. K., 3 *Bn*
(2) Fitzgerald, C., 3 *Bn.*
1Whittall, G. E. 8Jan.15
c.o. *Portal, J. L.* 17May15
2*Crosse, R. B.*, D.S.O., *Adjt.* 17May15
Boyle, C. R. C. 17May15
s. *Terry, W. E. C.* 10June15
Tolson, W. G. 10June15
5*Paget, B. C. T.* 10June15
Powell, J. J. 10June15
(5) *Plowden, G. F.* 10June15
Davenport, F. M. 10June15
Courtis, J. H. 10June15
Southey, J. A. 10June15
Mundey, S. C. B. 10June15

Lieutenants.

s. *Paris, A. C. M.* (*Empld. under Admy.*) 14Aug.13
2Baines, C. S., *D.S.O.*, 8Oct.13
2Brett, R. J. 25Oct.13
1Wynter, F. C. W. 22Mar.14
2Owen, R. M. (*temp. capt.* 22*June*15) 18Apr.14
1Birch Reynardson, H. T. 25July14
(9) Hanbury-Williams, C. F. R. 6Aug.14
(2) Wingfield, M. ff. R., 3 *Bn.* —
2Boardman, J. H. 17Sept.14
1Horan, K. 17Sept.14
(2) Warner, C. J., 3 *Bn.* —
(2) Fowke, C. A. F., 3 *Bn.* —
2Titherington, G. W. 8Jan.15
1Kearsley, J. S. 28Feb.15
2Barran, C. A. 17May15

2nd Lieutenants.

2Withington, T. E. 23Jan.14
*2Newton-King, P. F. [L] 25Feb.14
(2) Chippindale, F. W. C., *Spec. Res.* —
2*Sewell, D. A. D.* 8Aug.14
*2Rendel, R. D. 12Aug.14
(2) *Wyld, J. W. G. 15Aug.14
(2) *Whitfeld, E. H. 15Aug.14
†Baillie, C. W. H. (*attd. Rif. Brig.*) 15Aug.14
(2) Johnston, J. L., 3 *Bn.*
(2) Minifie-Hawkins, S. M., 3 *Bn.*
(2) Chevallier, C. T., 3 *Bn.*
(2) Ridout, P. M., 3 *Bn.*
(2) Carew Hunt, A. N., 3 *Bn.*
(2) Webster, P. L. C., 3 *Bn.*
2Meade, J. W. (*temp. lt.* 17 *May* 15) 1Oct.14
2Cardy, A. G. (*temp. lt.* 17 *May* 15) 17Oct.14
(2) Eighteen, W. J. 4Nov.14
1Mason, A. E. 7Nov.14

2nd Lieutenants—contd.

Loverock, R. C. 7Nov.14
1Naylor, G. 7Nov.14
Murphy, D. 7Nov.14
Brown, F. 7Nov.14
2Field, G. 9Nov.14
2Dean, C. 9Nov.14
2Wooding, W. W. 9Nov.14
2Kite, R. B. 11Nov.14
2Barnard, W. L. 11Nov.14
2Belgrave, J. D. 16Dec.14
2Hurst-Brown, C. 16Dec.14
2Littledale, W. J. 16Dec.14
2Hughes E. R. G. 23Dec.14
2Baker, C. B. (*temp.*) 1Jan.15
2 Martin, V. C. (*temp.*) (*attd. Rif. Brig.*) 1Jan.15
(3) Slade-Baker, J. B. 13Jan.15
(3) Clarke, N. G. 13Jan.15
2Solomon, J. B. (*temp.*) 27Jan15
2Horley, C. R. (*temp.*) 27Jan15
(2) Pendavis, H. V., *D.S.O.* 31Jan.15
2Peploe, K. 17Feb.15
2Grant, J. G. 6Mar.15
(3) Fanning, V. E. 16June15
(3) Smyth, H. E. F. 16June15
(3) Tyrwhitt-Drake, T. 14July15
(3) Hardcastle, J. B. 11Aug.15
(3) Sturges, O. H. M. 11Aug.15

Adjutants.

2Crosse, R. B., *D.S.O.*, *capt.* 1Mar.13
1Foljambe, *Hon.* J. C. W. S., *capt.* 6May13

Quarter-Masters.

r. *King, T.* 24Nov.97 *hon. maj.* 24Nov.12
1Ivey, T. 22Sept.03 *hon. capt.* 22Sept.13
5Roberts, F. M., *hon. lt.* 25Aug.14
2Warnock, A., *hon. lt.* 19Mar.15

*Temporary lieutenant 15 Nov. 14.

† Temporary lieutenant 8 Jan. 15.

THE OXFORDSHIRE AND BUCKINGHAMSHIRE LIGHT INFANTRY—(Regtl. Dist. No. 43)—*contd.*

Special Reserve.

Lieutenants.

MacKeown, S. E. (*attd. Rif. Brig.*) 12Dec.14
Button, G. T. (*H*) 10Dec.14

2nd Lieutenant.

(2) Chippindale, F. W. C. 5Aug.14

3rd Battalion. (Reserve.)

(*See page xl. as to honorary Army rank granted on account of Militia embodiment.*)

Hon. Colonel.

s. ✠*Marlborough*, Rt. Hon. *C. R. J.*, Duke of, K.G., TD. (*Lt.-Col. ret. T.F.*) (*temp. Lt.-Col. in Army*) 14June08

Lt.-Colonel.

Higgins Bernard, F. T. (Q) (*H*) 31Mar.12

Major.

p.s. Blyth, J., *hon. l.c.* (*H*) 5May06
s. ✠Paske, G. F., *hon. l.c.* (*Hon. Capt. in Army* 2*Nov.*00) (*H*) 12Sept.06
r.e. ✠Darell-Brown, H. F. 18Sept.07

Captains.

s. ✠*Fitzwilliam, W. C. de M.*, Earl, K.C.V.O., D.S.O., *hon. m.* (*Bt. Lt.-Col. W. Rid. R.H.A.*) 11Apr.96
p.s. Barrington, *Hon.* W. R. S., *hon. m.* (*H*) 6Jan.97
(2) North, A. K., *late* 2nd Lt. A.S.C. 4June02
Godsal, P. (*H*) 7June02
✠*Cardwell, C. A.*, Capt. ret. pay (*H*) Ⓢ 7Apr.06
s. *Salkeld, R. E.* (*Dist. Commr., E. Africa Prote.* 12*July*06) 12July06
p.s. *Dashwood, R. H. S.*, Capt. ret. pay (*H*) Ⓢ s. 1Dec.07
(2) Fitzgerald, C., Lt. ret pay 8May11
Jackson, H. N., Lt. ret. pay (*Instnl. Duties*) 14Aug.13
d. Maul, H. C. (*H*) 10Oct.14
p.s. Tatton, R. H. G. (*H*) (*attd. D. of Corn. L.I.*) 10Oct.14

Lieutenants.

✠Spencer, A. V., *D.S.O.* (*H*) 5Mar.10
(2) Wingfield, M. ff. R., *late* 2nd Lt. Oxf. & Bucks. L.I. 27Aug.14
(2) Warner, C. J. 26Sept.14
Smith, J. N. (*Asst. Dist. Offr., N. Provs., Nigeria*, {1*Jan.*14 / 19*Sept.*10) 29Oct.14
(2) Fowke, C. A. F. 10Dec.14
Evetts, A. F. 24Apr.15
Rawson, H. W. H. 10May15

2nd Lieutenants.

Tuck, D. J. (*attd. D. of Corn. L.I.*) 8Aug.14
(2) Johnston, J. L. 15Aug.14
Newbolt, A. F. (*attd. D. of Corn. L.I.*) 15Aug.14

2nd Lieutenants—contd.

Goodwyn, L. J. (*attd. Yorks. L.I.*) 15Aug.14
Quinn, J. W. (*attd. D. of Corn. L.I.*) 15Aug.14
(2) Minifie-Hawkins, S. M. 15Aug.14
Walters, F. P. (*attd. Yorks L.I.*) 15Aug.14
(2) Chevallier, C. T. 15Aug.14
(2) Ridout, P. M. 15Aug.14
(2) Carew Hunt, A. N. 15Aug.14
(2) Webster, P. L. C. 15Aug.14
Wood, A. W. 15Aug.14
Vidal, L. A. (*attd. D. of Corn. L.I.*) 15Aug.14
Slocock, C. H. B. (*on prob.*) 2Dec.14
Williams, T. P. (*on prob.*) 16Dec.14
Radford, H. D. H. (*attd. E. Kent R.*) 30Dec.14
McConnell, R. L. (*attd. Rif. Brig.*) 30Dec.14
Griffith, E. N. (*on prob.*) 2Jan.15
Dashwood, H. G. (*attd. Rif. Brig.*) 3Jan.15
w.a. *Rowe W. C.* (*on prob.*) 27Feb.15
Riddle, A. E. S. (*attd. E. Lanc. R.*) 30Mar.15
Foreshew, C. E. P. (*on prob.*) 31Mar.15
Foreshew, T. W. C. (*on prob.*) 31Mar.15
Griffith-Williams, M. S. *on prob.*) 8Apr.15
Strickland, R. (*on prob.*) 9Apr.15
Stephens C. G. (*on prob.*) 11Apr.15
Widcombe, C. I. (*on prob.*) 17Apr.15
Huggard, G. C. (*on prob.*) 18Apr.15
Lawrence, W. G. (*on prob.*) 20Apr.15
Atkins, C. (*on prob.*) 23Apr.15
Griffin, S. J. (*on prob.*) 23Apr.15
Coles, W. T. (*on prob.*) 23Apr.15
FitzGerald, R. A. (*on prob.*) 25Apr.15
Truman, A. H. (*on prob.*) 4May15
Gardner, A. E. (*on prob.*) 5May15
Webster-Jones, A. O. W. (*on prob.*) 9May15
Adams, H. C. (*on prob.*) 14May15
Brett, F. S. (*on prob.*) 16May15
Riley, C. H. (*on prob.*) 19May15
Rance, W. (*on prob.*) 20May15
May, R. A. (*on prob.*) 26May15
Davenport, C. T. (*on prob.*) 29May15
Tunnard, T. E. (*on prob.*) 29May15
§Kellet, J. A. 10June15
§Davis, A. H. 11June15
§Watts, L. R. 11June15
§Milford, T. R. 11June15
§Campbell, H. D. C. 14June15
§Seccombe, P. 14June15
§Fussell, J. W. H. 14June15
§Rock, L. J. 17June15
§Holiday, W. G. 24June15

2nd Lieutenants—contd.

§Crocombe, F. R. 26June15
§Roberts, B. F. 26June15
§Warde, B. C. C. 26June15
§Reddy, T. H. 26June15
§Ionides, T. A. 14July15
§Scott, J. H. F. 24July15
§Lester-Smith, W. C. 24July15
§Rose, J. J. T. 24July15
§Walden, H. R. 31July15
§Line, W. 4Aug.15
§Owen, L. V. D. 26Aug.15
§England, R. 26Aug.15
§Hutchison, T. 26Aug.15
§Osborne, V. E. 26Aug.15

Adjutant.

✠Frith, C. H., Capt. Oxf. & Bucks. L.I. 17Sept.12 (*Capt. in Army* 22*Nov.*02)

Quarter-Master.

✠King, T., *hon. maj.*

4th Battalion. (Territorial.)

"South Africa, 1900-01."
St. Cross Road, Oxford.

Hon. Colonel.

North, W. H. J., Lord, TD, *late* Lt. 1 L.G. 9Jan.95

Lt.-Colonels

2p.s. Ames, W. H., TD. (Q), *hon. c.* *8Sept.14 11June04
3p. Stockton, A., TD, (*H*) Ⓢ *30Apr.15 30Nov.09
1✠Dugmore, W. F. B. R., *D.S.O.*, Capt. ret. pay (*Res. of Off.*) *3May15

Majors.

1p.s. Schofield, F. W. (*H*) (Q) (**Lt.-Col.* 1 *Jan.* 15), s. 1Apr.08
1p. Ovey, R. L. 30Nov.09
2Beaman, G. P. R. Maj. ret. Ind. Army *27Sept.14
2✠Slessor, A. K., Capt. ret. pay, *Adjt.* *29Sept.14
2Seymour, G. W., *late* Capt. R. Ir. Fus. *30Sept.14

Captains.

p.s.1Rowell, R. R. S. (**Maj.* 18 *Dec.* 14) 20Dec.15
1p.s. Rose, D. M. (*H*) 20Dec.05
1p.s. Fortescue, E. C. 1Apr.08
c.p.s. Fox, T. S. W. (*H*) 1Dec.09
1Conybeare, J. J. *5Aug.14 4Dec.12
1p. Coleman, E. G. *5Aug.14
1Hadden, E. W. R. 2Sept.14

Lieutenants.

Twisleton-Wykeham-Fiennes, Hon. *L. J. E.*, *f.c.* (**Capt.* 27*Oct.*14) 1June13
1Jones, F. B. (**Capt.* 13 *May* 15) 1Jan.14
Long, B. (*H*) (**Capt.* 15 *May* 15) (*Brig. M.G. Offr.*) 1Jan.14
1Pickford, P. (**Capt.* 10 *June* 15) 1Jan.14
1Treble, J. N. (**Capt.* 28 *Oct.* 14) 21Jan.14
2Brucker, A. H. (**Capt.* 9 *Dec.* 14) 14Sept.14
2Gray, W. T., *late* Lt. Impl. Yeo. (**Capt.* 9*Dec.* 14) 14Sept.14

Lieutenants—contd.

2Bennett, H. J. (**Capt.* 9 *Dec.* 14) 14Sept.14
2Davenport, H. N. (**Capt.* 9 *Dec.* 14) 14Sept.14
2Cuthbert, R. F. (**Capt.* 9 *Dec.* 14) 14Sept.14

2nd Lieutenants.

Taylor, F. W. (*Jun. Supt., Educn. Dept., Nigeria* 17*June*14) 25July12
2*Jee, A. C.* (**Capt.* 9 *Dec.* 14) 1May13
1Grisewood, A. A. (**Lt.* 9 *Jan.* 15) 1Nov.13
3Barton, C. J. (**Lt.* 13 *July* 15) 1Apr.14
1Rose, G. K. (**Capt.* 10 *June* 15) 6Aug.14
1Griffin, I. E. (**Lt.* 9 *Jan.* 15) 8Aug.14
2Abraham, R. L. (**Lt.* 28 *July*15) 27Aug.14
1Deacon, H. J. (**Lt.* 9 *Jan.* 15) 2Sept.14
1Cranmer, J. E. A. 2Sept.14
2Stockton, J. G. (**Lt.* 26*July* 15) 7Sept.14
1Wilsdon, H. A. (**Lt.* 13 *May* 15) 2Sept.14
1Waymann, W. A. 2Sept.14
1Brooks, B. B. B. 2Sept.14
1Gibson, A. K. (**Lt.* 10 *June* 15) 12Sept.14
1Grisewood, F. H. (**Lt.* 10*June*15) 12Sept.14
Barnett, J. C. L., s. (**Lt.* 9 *Dec.* 14) 14Sept.14
2Richardson, M. C. 14Sept.14
2Davis, A. G. A. (**Lt.* 9 *Dec.* 14) 14Sept.14
2Boyle, R. F. R. P. (**Capt.* 23 *July* 15) 14Sept.14
2Doyne, P. D. (**Capt.* 9 *Dec.* 14) 14Sept.14
2Marcon, C. S. W. (**Lt.* 25 *July* 15) 14Sept.14
2Harris, H. T. T. (**Capt.* 9 *Dec.* 14) 14Sept.14
2Bridges, E. E. (**Lt.* 27 *July* 15) 14Sept.14
2Smith, B. H. (**Lt.* 9 *Dec.* 14) 14Sept.14
Brown, D. C. 14Sept.14
2Brown, K. E. 14Sept.14
2Sanderson, J. F. B. 6Oct.14
1Greenwell, G. H. 8Oct.14
3Andrews, A. N. (**Capt.* 5 *June* 15) 8Oct.14
1Edmunds, M. W. 8Oct.14
2Scott, W. D. (**Lt.* 23 *July* 15) 11Jan.15
1Cooper, M. C. 16Jan.15
2Simpson-Hayward, G. H. (**Lt.* 24*July*15) 20Jan.15
2Parsons, C. R. 20Jan.15
2*Strang, M.* 21Jan.15
1Fortescue, T. R. 23Jan.15
2Hall, T. N. 1Feb.15
2Etty, J. L. 1Feb.15
2Tiddy, R. J. E. (**Lt.* 29 *July* 15) 16Feb.15
2Procter, A. W. 19Feb.15
2*Mitchell, J. G.* 26Feb.15
1Craig, C. C. 5Mar.15
1Judson, H. L. 9Mar.15
1Lake, R. St. G. 9Mar.15
1Mason, C. R. 19Mar.15

§ On probation

THE OXFORDSHIRE AND BUCKINGHAMSHIRE LIGHT INFANTRY—(Regtl. Dist. No. 43)—*contd.*

4th Bn.—*contd.*

***2nd Lieutenants*—contd.**

3Wood, G. R. (**Capt.* 1 *May* 15) 1May15
3Hunt, A. S. (**Lt.* 1 *May* 15) 1May15
2Zeder, J. H. 1May15
3Stockton, H. O. 3May15
3Gamlen, J. C. B. 6May15
3Stevenson, G. H. (**Lt.* 6 *May* 15) 6May15
3Hunter, L. W. 6May15
3Lakin, C. 6May15
3Stainer, C. L. (**Lt.* 6 *May* 15) 6May15
2Greey, F. L. 12May15
2Shepherd, G. H. G. 15May15
2Shepherd, J. G. 15May15
2Hunt, R. N. C. 19May15
3Wallace, W. J. L. 1June15
2Morden-Wright, H. 10June15
Palmer, G. H. E. 12June15
2Miller, J. G. R. 19June15
2Smith, E. E. 5July15
Enoch, W. H. 15July15
2Moberley, W. H. 17July15 1May15
Wrong, H. H. 20July15
Callender, J. C. 23July15
Searby, H. R. 23July15
Blake, J. E. 24July15
2Coombs, H. E. 24July15
Rawlinson, G. M. 4Aug.15

Adjutants.

1✕Ballard, J. A., Capt. Oxf. & Bucks L.I. 9May14
3✕Slessor, A. K., *maj.* 27Apr.15

Quarter-Masters.

1p.s.✕Bridgewater, A. A., *hon. m.* 30Dec.05
2Hobbs, W. A., *hon. lt.* 24Nov.14
3King, W. G., *hon. lt.* 1May15

Medical Officers.

1Summerhayes, Maj. J. O., R.A.M.C. (T.F.) (*attd.*) 25Apr.15 8Feb.13
2Howard, Capt. V., R.A.M.C. (T.F.) (**Maj.* 17 *Jan.* 15) (*attd.*) 5Aug.14 29May12
Meikle, Capt. R. W., R.A.M.C. (T.F.) (*attd.*) 14Sept.14
Gauntlett, Lt. H. L., R.A.M.C. (T.F.) (*attd.*) 1Apr.13

Chaplains.

Cary-Elwes, *Rev.* A., *M.A.*, Chapl. 4th Class (T.F.) (*attd.*) 1Apr.08 14Mar.06
Smith, *Rev.* E. F., *M.A.*, Chapl. 4th Class (T.F.) (*attd.*) 1Apr.08 28Mar.06
Streatfeild, *Rev.* F., Chapl. 4th Class (T.F.) (*attd.*) 10Oct.14

[**Uniform—*Scarlet.* Facings—*White.***]

Cadet Units affiliated.
Cowley Cadet Corps.
Burford Grammar School Cadet Corps.

Buckinghamshire Battalion. (Territorial).

"South Africa, 1900-02."
Aylesbury.

Hon. Colonel.

Lt.-Colonels

p.s. *Fremantle,* Hon. *T. F.*, VD (*T*) (*H*), *s.* 24Mar.06
2p. Williams, H. M., VD. (*t*) *14Aug.14 18Apr.11
1Doig, C. P. *9Mar.15

Majors.

C. Hooker, J. H. 17Oct.11
2Chadwick, J., TD. 18Apr.14
3Gilbey, A., VD. (*Lt.-Col. and Hon. Col. ret. Vols.*) (*to command Bn.*) *11Apr.15

Captains.

1✕Hawkins, L. C. (*Hon. Lt. in Army* 18*June* 01)(**Maj.* 1*Sept* 14) 31Aug.01
p.s. Reynolds, L. L. C. 18July03
2p. Barrett, H. L. C. 19Mar 04
Baker, A. B. L., *s.* 19June12
1p.s. Crouch, L. W. (Q)(*H*) 19June12
3p.s. Addington, J. G., *Lord* (*H*) 20Sept.13
1Birchall, E. V. D. 5Oct.13
Mitchell, C. J., Capt. Res. of Off., *s.* (**Maj.* 1 *Sept.* 14) 18 Apr.14
2p.s. Christie-Miller, G. *5Aug.14 5Aug.07
p.s. *Vernon, S. R.* (*Capt. ret. T.F.*) *s.* *5Aug.14 21Jan.11
3Bowyer, G. E. W. 1Sept.14
2p. Hankin, G. T. (q) *3Nov.14 17July.08
✕*Harling, R. W.*, Capt. ret. pay, (*Res. of Off.*) *s.* *11Nov.14
3✕Pembroke, A. B., Capt., Res. of Off. (*Hon. Maj. ret. Spec. Res.*) (*Hon. Capt. in Army* 11 *June*01) *Adjt.* *14Apr.15

Lieutenants.

Littledale, E. H. (*temp. Lt. R.E.*) 1July11
1Crouch, G. R. ⓢ (**Capt.* 1 *Sept.* 14) 19June12
1Hall, P. A. (*H*) (**Capt.* 1 *Sept.* 14) 17Sept.12
1Reid, N. S. (**Capt.* 12 *Sept.* 14) 17Sept.12
1Jackson, G. G. (**Capt.* 1 *Sept.* 14) 6Nov.13
1Combs, H. V. (**Capt.* 21 *May* 14) 1Feb.14
1(p) Vernon, A. S. (**Capt.* 6 *June* 15) *5Aug.14 12Sept.02
1Viney, O. V., (**Capt.* 6 *June* 15) 1Sept.14
C. Buckmaster, R. A. 2Sept.14
2Fremantle, *Hon.* W. *21Sept.14

2nd Lieutenants.

1Green, B. (**Lt.* 1 *Sept.* 14) 24June14
2Simpson, G. H. (**Capt.* 5 *May* 15) 22July14
2*List, A. P.* (**Capt.* 10 *Oct.* 14) 5Aug.14
1Kennish, A. C. E. F. (**Lt.* 28 *Feb.* 15) 5Aug.14
1Backhouse, J. W. (**Capt.* 10 *June* 15) 1Sept.14
1Reynolds, F. G. B. (**Lt.* 21 *Apr.* 15) 1Sept.14
Bowen, M. 1Sept.14
2Wyllie, G. A. 2Sept.14
1Norwood, R. C. 2Sept.14
1Brown, A. D. B. (**Lt.* 6 *May* 15) 2Sept 14
1Young, R. E. M. 2Sept 14
1Woollerton, E. N. C. (**Lt.* 6 *May* 15) 2Sept.14
1Hobert-Hampden, G. M. A. (**Lt.* 6 *May* 15) 2Sept.14
1Wright, E. L. (**Capt.* 10 *June* 15) 3Sept.14
1Earl, F. D. (**Lt.* 1 *July* 15) 5Sept.14
2Stewart Liberty, I. (**Capt.* 25 *July* 15) 6Sept.14
1Hill, J. S. B. (**Lt.* 10 *Sept.* 14) 10Sept.14
2Stevens, G. C. (**Lt.* 16 *Aug.* 15) 21Sept.14
2Buckmaster, H. S. G. (**Capt.* 5 *May* 15) 21Sept.14
2Rolleston, J. M. (**Lt.* 20 *Aug.* 15) 21Sept.14
2Hughes, R. T. (**Lt.* 19 *Aug.* 15) 21Sept.14
1Gregson-Ellis, R. G. (**Lt.* 10 *June* 15) 21Sept.14
2Cummins, W. A. (**Lt.* 15 *Aug.* 15) 30Sept.14
1Troutbeck, G. L. 20Oct.14
2Ranger, V. W. H. (**Lt.* 17 *Aug.* 15) 20Oct.14
2Phipps, C. P. 31Oct.14
2Chadwick, D. G. (**Lt.* 18 *Aug.* 15) 31Oct.14
2Firminger, J. E. 3 Oct.14
2Church, H. (**Capt.* 25 *July* 15) 31Oct.14
2Greene, W. A. (**Capt.* 22 *Aug.* 15) 2Dec.14
2Symonds, R. F. 24Dec.14
1Wright, P. L. (**Lt.* 10 *June* 15) 4Feb.15
1Neave, G. V. 18Feb.15
1Hales, J. B. 20Feb.15
2Letts, E. M. 20Feb.15
3Bishop, S. D. 20Feb.15
2Atkinson, G. W. 20Feb.15
2Williamson, N. B. 20Feb.15
2Newbery, B. J. 19Mar.15
1Pullman, H. J. 23Mar.15
2Fordham, O. 23Mar.15
3Gilbey, A. R. (**Lt.* 28 *July* 15) 5May15 22Jan.15
Gray, J. McC. T. 15May15
3Chapman, A. F. 31May15
3Chapman, J. P. 1June15
3Godfrey, A. P. 2June15
3Rowland, R. 3June15
2Wells, P. E. 9June15
3Hall, C. 18June15
3Irby, G. N. 19June15
2Long, E. W. (**Lt.* 20 *June* 15) 20June15
2Hudson, R. B. 26June15

***2nd Lieutenants*—contd.**

2Vaughan, F. G. 17July15
Heath, W. R. 19July15
2Sanford, J. C. 21July15
3Furley, R. B. 23July15
Bates, E. G. H. 27July15
Foster, H. R. 1Aug.15
Flint, N. S. 14Aug.15
Quayle, C. P. 22Aug.15

Adjutants.

1Bartlett, A. J. N., Capt. Oxf. & Bucks. L. I. 25Feb.13
✕Pembroke, A. B. *Capt.* 14Apr.15

Quarter-Masters

2Waller, D., *hon. lt.* 25Sept.14
1✕Nichol, E., *hon. lt.* 17Feb.15
3Inder, H. J., *hon. lt.* 8May15

Medical Officers.

2p. Baker, J. C., *Surg.-Capt.* 23Oct.07 23Apr.04

Chaplains.

Wethered, *Rev.* O. H., *M.A.*, TD, Chapl. 2nd Class (T.F.) (*attd.*) 2June12 2June97
Whitechurch, *Rev.* V. L., Chapl. 4th Class (T.F.) (*attd.*) 2Dec 14
Phipps, *Rev. Canon* C. O., *M.A.*, Chapl. 4th Class (T.F.) (*attd.*) 2Dec.14

[**Uniform—*Dark Grey* Facings—*Scarlet.***]

Cadet Units affiliated.
Aylesbury Grammar School Cadet Corps.
Slough Secondary School Cadets.

5th (Service) Battalion.

In Command.

Cobb, Maj. (*temp. Lt.-Col.*) C. H., Oxf. and Bucks. L.I. 19Aug.14 29Nov.00

Major.

(***2nd in Command.***)

Majors.

Webb, W. F. R. (*temp.*) (*Capt.* 22 *Punjabis* 13Mar.15

Captains.

Logan, Capt. R. O., Oxf. & Bucks. L.I. 21Aug.14 19Nov.10
Sanderson, Capt. A. E., Oxf. & Bucks. L.I. 25Aug.14 25July14
Carfrae, Capt. C. F. K., Oxf. & Bucks. L.I. 29Aug.14 5Aug.14
Webb, A. W. T. (*temp.*) (*Lt. Ind. Army*) 1Oct.14
Plowden, Capt. G. F., Oxf. & Bucks. L.I. 1Oct.14 10June11
Barwell, N. F. (*temp.*) 26Apr.1•
Cobb, H. H. (*temp.*) 26Apr.1•

1299 | 1299a | 1299b | 1300

THE OXFORDSHIRE AND BUCKINGHAMSHIRE LIGHT INFANTRY–(Regtl. Dist. No. 43)–*contd.*

5th Bn.—*contd.*

Lieutenants.

Maude, E. W. (*temp.*) 1Oct.14
Cooke, B. K. (*temp.*) 1Oct.14
Birch, W. R. (*temp.*) 1Oct.14
Crawford, C. B. (*temp.*) 1Oct.14
Ridge-Jones, A. (*temp.*) 7Oct.14
Day, H. J. T., 9 Bn.
Jury, C. R. (*temp.*) 15Dec.14 18June15
Bird, S. H., 9 Bn.
Newman, L.S.M., 9 Bn.
⚔Walker, A. C. (*temp.*) 23Apr.15 18June15
Sweet-Escott, L. W. (*temp.*) 13May15
Clarke, H. F. (*temp.*) 20May15

2nd Lieutenants.

Fenwick, J. S. (*temp.*) 22Aug.14
Lee, L. S. (*temp.*) 26Aug.14
Beckingham, W.T. (*temp.*) 22Sept.14
Fremantle, T.F.H. (*temp.*) 30Sept.14
Walter, C. H. (*temp.*) 10Oct.14
Curry, P. A. (*temp.*) 2Dec.14
Bleeck, J.(*temp.*)4Dec.14
Le Messurier, H.A. (*temp.*) 30Dec.14
Carter, K. V. (*temp.*) 10Mar.15
Cupper, H. J. (*temp.*) 20Mar.15

Adjutant.

Paget, Capt. B. C. T., Oxf. & Bucks. L.I. 14Sept.14 (*Temp. Capt. in Army* 1 *Oct.* 14)

Quarter-Master.

Roberts, F. M., *hon. lt.* 25Aug.14

6th (Service) Battalion.

In Command.

White, Lt.-Col. E. D. 6Sept.14 23Oct.11

Major.
(*2nd in Command.*)

Childers, Maj. E. M., Res. of Off. 5Sept 14

Major.

Davy, Maj. J. D. W., Res. of Off. 22Oct.14
Foljambe, *Hon.* G. W. F. S. (*temp.*) (*Capt. Res. of Off.*) 22Oct.14

Captains.

Vyner, R. J. (*temp.*) 23Sept.14
Williams, R. M. (*temp.*) 11Oct.14
Osborne, J. E. (*temp.*) 22Oct.14
Pears, T. (*temp.*)5Jan.15
Bryant, J. E. (*temp.*) 18Jan.15
Stephens, H. E. (*temp.*) 5Feb.15

Lieutenants.

Anderson, E. J. (*temp.*) 22Oct.14
Milne, J. T. (*temp.*) 30Dec.14
Boissier, J. S. (*temp.*) 30Dec.14
Baines, A. B. (*temp.*) 5Feb.15
Smith, H. V. (*temp.*) 5Feb.15
Hunt, R. L. G. (*temp.*) 5Feb.15
Shaw, E. A. (*temp.*) 5Feb.15
Lack, V. J. F. (*temp.*) 5Feb.15
Middleditch, G. E. (*temp.*) 26Mar.15
Johnston, C. F. (*temp.*) 5July15

2nd Lieutenants.

Rogers, R. B. (*temp.*) 19Sept.14
Fagan, B. W. (*temp.*) 13Nov.14
Sutherland, H J. (*temp.*) 16Nov.14
Gosset, A. C. V. (*temp.*) 23Nov.14
Burn, D. C. (*temp.*) 27Nov.14
Whitlock, F. W. (*temp.*) 2Dec.14
Goodwyn, C. C. (*temp.*) 22Dec.14
Boissier, G. D. (*temp.*) 29Dec.14
McIlroy, W. E. C. (*temp.*) 25Jan.15
Tubbs, A. (*temp.*) 9Apr.15

Adjutant.

Quarter-Master.

Brazier, W. *hon. lt.* (*temp.*) 22Sept.14

7th (Service) Battalion.

In Command.

Newton-King, Lt.-Col. (*temp.*) F. J. 22Jan.15

Major.
(*2nd in Command.*)

Wheeler, C. (*temp.*) 11July15 9Feb.15

Majors.

Debenham, F. (*temp.*) 11July15

Captains.

⚔Martin, G. B. (*temp.*) 17Dec.14
Ellis, L. J. (*temp.*) *Adjt.* 1Jan.15
⚔Guise, J. (*temp.*)15Feb.15
Royal-Dawson, O. S. (*temp.*) 11June15
Salvesen, C. A. (*temp.*) 10July15
Arnett, M. G. (*temp.*) 10July15
Simpson, B. L. (*temp.*) 10July15

Lieutenants.

Pemberton, R. B. (*temp.*) 8Nov.14
Wicks, C. L.(*temp.*) 17Dec.14
Miller, G.C.(*temp.*) 17Dec.14
Stahlschmidt, F.A D. (*temp.*) 17Dec.14

Lieutenants—contd.

Henry, H. A. (*temp.*) 2June15
Peirson, N. J. (*temp.*) 12June15
Ditchburn, A. H. (*temp.*) 8July15
Keble, J. H. C. (*temp.*) 10July15
Baker, C. P. (*temp.*) 11July15

2nd Lieutenants.

Bowman, T. (*temp.*) 22Sept.14
Streatfeild, R. C. (*temp.*) 22Sept.14
Ker, C. P. (*temp.*) 24Oct.14
Symonds, S. L. (*temp.*) 14Nov.14
Garland, W. (*temp.*) 24Nov.14
Riley, E. (*temp.*) 27Nov.14
Northcote, D. S. (*temp.*) 27Nov.14
Hathaway, T. W. (*temp.*) 28Nov.14
Hoey, J. T. S. (*temp.*) 7Dec.14
Young, W. A. (*temp.*) 18Dec.14
Thomas, A. L. (*temp.*) 24Dec.14
Boor, A. P. (*temp.*) 28Dec.14
Andrews, P. E. (*temp.*) 3Jan.15
Molyneux, P. L. (*temp.*) 11Jan.15
Green, C. H. (*temp.*) 25Jan.15
Manning, C. A. W. 25Jan.15
Walding, T. W. (*temp.*) 25Jan.15
Blake, W. T. (*temp.*) 25Jan.15
Shaw, J. W. (*temp.*) 27Mar.15
Holiday, A. S. (*temp.*) 1Apr.15
Robinson, H. W. (*temp.*) 7Apr.15
Riley, G. W. (*temp.*) 9Apr.15
Bobby, A. S. (*temp.*) 28Apr.15

Adjutant.

Ellis, Capt. (*temp.*) L. J. 1Jan.15

Quarter-Masters.

⚔Powell, T., *hon. lt.* (*temp.*) 2Jan.15

8th (Service) Battalion. (Pioneers.)

In Command.

Justice, Bt. Col. C. Le G., Ind. Army 19Sept.14 11May07

Major.
(*2nd in Command.*)

Llewellyn, F. (*temp.*) 30Oct.14

Major.

Darnley, R. (*temp.*) 10May15

Captains.

Burt, A. E. (*temp*) 14Nov.14
Lowe, A. H. (*temp.*) 14Nov.14
Tardrew, C. G. (*temp.*) 1Mar.15
Dunthorne, R. G. (*temp.* 1Mar.15
Eales, T. F. J. (*temp.*) 1Mar.15
Bartrum, G. (*temp.*) 10May15
Garden, C. A. (*temp.*) 11May15

Lieutenants.

Leman, S. C. (*temp.*) 14Nov.14
Thompson, H. W. A. (*temp.*) 14Nov.14
Payne, E. H (*temp.*) 1Mar.15
Edwards, A. R. (*temp.*) 1Mar.15
Forwood, L. L. (*temp.*) 1Mar.15
Morris, G. P. (*temp.*) 1Mar.15
Maul, G. B. (*temp.*) 1Mar.15
Groocock, H. L. (*temp.*) 15Mar.15
Giffard, W. L. (*temp.*) *Adjt.* 1May15
Rayner, C. S. W. (*temp.*) 10May15
Medcalf, E. F. (*temp.*) 11May15

2nd Lieutenants.

Straight, F. S. (*temp.*) 5Nov.14
Calder, A. (*temp.*) 13Nov.14
Benjamin, A. F. (*temp.*) 6Dec.14
Haggen, G. L. (*temp.*) 7Dec.14
Amery, G. D. (*temp.*) 10Dec.14
Elliott, N. W. (*temp.*) 23Dec.14
Pierce, W. M. (*temp.*) 7Jan.15
Smith, J. E. P. (*temp.*) 7Jan.15
Watts-Watts, T. N. (*temp.*) 9Jan.15
Jee, R. (*temp.*) 11Jan.15
Bartrum, G. (*temp.*) 22Jan.15
Brimblecombe, P. Y. (*temp.*) 29Jan.15
Rose, G. S. (*temp.*) 24Feb.15
Edwards, A. H. (*temp.*) 26Feb.15
Sheppard, C. W. (*temp.*) 1Mar.15
Curtis, H. E. (*temp.*) 10Mar.15
Shaddock, W. H. (*temp.*) 13Apr.15
Taafe, C.(*temp.*)13July15

Adjutant.

Giffard, Lt. (*temp.*) W. L. 1May15

Quarter-Master.

Woollen, W., *hon. lt.* (*temp.*) 9Nov.14

THE OXFORDSHIRE AND BUCKINGHAMSHIRE LIGHT INFANTRY—(Regtl. Dist. No. 43)—*contd.*

9th (Reserve) Battalion.

In Command.

✕Dyas, Bt. Col. J.R., p.s.c. [l] 7Aug.15 22Dec.08

Major.

(*2nd in Command*) ✕Benson, C. B., *D.S.O.*, (*temp.*) (*Capt. Res. of Off.*) [L] 28Dec.14

Major

Crum, A. S. (*temp.*) (*Hon. Capt.inArmy 25Jan.01*) 6May15 1Oct.14

Captains.

Williams, C. H. (*temp.*) 28Dec.14
Hogan, C. J. (*temp.*) 28Dec.14
Knighton, G. G. (*temp.*) 28Dec.14
Skinner, J. H. McI. (*temp.*) 28Dec.14
Inman, H. J. (*temp*). 7Jan.15
Evans, A. G. (*temp.*) 14Jan.15
Labouchere, A. M. (*temp.*) 20Jan.15

Lieutenants

Hanbury-Williams, Lt. C. F. R., Oxf. & Bucks. L.I. — 6Aug.14
(5) Day, H. J. T. (*temp.*) 3Nov.14
Lardner, G. S. (*temp.*) 1Dec.14
Jacob, V. V. (*temp.*) 14Dec.14
(5) Bird, S. H. (*temp.*) 28Dec.14
Foot, E. L. (*temp.*) 28Dec.14
Calderon, G. (*temp.*) 9Jan.15
Rattray, S. (*temp.*) 18Jan.15
(5) Newman, L. S. M. 19Jan.15
Higgins, D. S. (*temp.*) 2Feb.15
Davies, G. H. (*temp.*) 18Mar.15

2nd Lieutenants.

Melliss, A. P. J. (*temp.*) 22Sept.14
Bowman, J. (*temp.*) 22Sept.14
Statt, C. C. (*temp.*) 22Sept.14
Chown, L. G. (*temp.*) 5Oct.14
Bousfield, G. W. J. (*temp.*) 23Oct.14
Paton, J. E. F. (*temp.*) 28Oct.14
Shadbolt, H. J. (*temp.*) 27Oct.14
Galloway, A. G. A. (*temp.*) 29Oct.14
Willock, R. P. (*temp.*) 9Nov.14
Ford, W. O. (*temp.*) 9Nov.14

2nd Lieutenants—contd.

Dick, J. G. (*temp.*) 19Nov.14
Morris, C. G. N. (*temp.*) 24Nov.14
de Pass, G. P. B. (*temp.*) 24Nov.14
Robinson, E. L. D. (*temp.*) 24Nov.14
Guy, G R. (*temp.*) 30Nov.14
Barnes, R. M. (*temp.*) 7Dec.14
Whitehead, H. M. (*temp.*) 7Dec.14
Davidson, M. C. (*temp.*) 9Dec.14
McConnell, R. L. (*temp.*) 11Dec.14
Simpkin, W. (*temp.*) 11Dec.14
Lyle, G. S. La W. (*temp.*) 14Dec.14
Allsopp, A. T. W. (*temp.*) 18Dec.14
Peel, R. (*temp.*) 18Dec.14
Gibbons, A. G. (*temp.*) 22Dec.14
Coales, J. L. (*temp.*) 22Dec.14
Higham, T. F. (*temp.*) 2Dec.14
Muriel, W. G. (*temp.*) 22Dec.14
Barnes, D. T. (*temp.*) 22Dec.14
Harrison, N. S. (*temp.*) 23Dec.14
Beaver, H. A. (*temp.*) 28Dec.14
Punch, J. S. C. (*temp.*) 5Jan.15
Spary, V. C. (*temp.*) 22Jan.15
Carr, H. J. (*temp.*) 24Jan.15
Haward, G. (*temp.*) 25Jan.15
Swift, H. L. (*temp.*) 25Jan.15
Cockell, G. B. (*temp.*) 26Jan.15
Thomas, T. (*temp.*) 26Jan.15
Bradley, G. (*temp.*) 26Jan.15
Scott, R. M. (*temp.*) 30Jan.15
Campbell, C. B. (*temp.*) 23Feb.15
Stukeley, A. T. W. (*temp.*) 26Feb.15
Parsons, F. S. (*temp.*) 26Feb.15
Chappell, E. F. (*temp.*) 6Mar.15
Hill, H. A. (*temp.*) 10Mar.15
Sims, H. V. (*temp.*) 10Mar.15
Greene, G. S. (*temp.*) 11Mar.15
Johnson, F. H. (*temp.*) 13Mar.15
Fuller, A. W. F. (*temp.*) 17Mar.15
Barrell, E. G. (*temp.*) 20Mar.15
Hutchins, W. T. (*temp.*) 25Mar.15

2nd Lieutenants—contd.

Coombes, J. C. (*temp.*) 7Apr.15
Frieake, G. M. (*temp.*) 7Apr.15
Bartlett, J. C. (*temp.*) 7Apr.15
Grosvenor, G. R. (*temp.*) 14Apr.15
Creswell, R. A. (*temp.*) 22Apr.15
Parkinson, E. B. (*temp.*) 22Apr.15
Allday, E. C. J. (*temp.*) 3May15
Norman, B. A. K. (*temp.*) 3May15
Cox, T. J. (*temp.*) 5May15
Henry, B. J. (*temp.*) 5May15
Bedford, D. E. (*temp.*) 6May15
Spurge, H. W. (*temp.*) 12May15
Ware, W. I. B. (*temp.*) 12May15
Best, J. H. (*temp.*) 12May15
Johnson, D. L. (*temp.*) 12May15
Ramsay, A. H. (*temp.*) 12May15
Field, H. T. C. (*temp.*) 12May15
Freeth, E. (*temp.*) 12May15
Garrard, G. F. (*temp.*) 12May15
Wilmot, D. A. T. (*temp.*) 12May15
Coulthard, E. F. (*temp.*) 13May15
Phillips, G. W. (*temp.*) 20May15
Cook, E. C. (*temp.*) 28May15
Wilsdon, R. G. (*temp.*) 28May15
Tees, E. S. L. (*temp.*) 3June15
Carew-Hughes, C. F. (*temp.*) 10June15
Hoole, S. W. (*temp.*) 10June15
Turner, A. G. (*temp.*) 10June15
Daye, A. J. (*temp.*) 10June15
Weston-Webb, H. (*temp.*) 10June15
Winder, J. F. E. (*temp.*) 10June15
Marsh, V. B. (*temp.*) 10June15
Middleditch, A. W. (*temp.*) 10June15
Mitchelson, A. (*temp.*) 10June15
Orr, H. W. (*temp.*) 10June15
Wilson, W. G. (*temp.*) 10June15
Wordsworth, O. B. (*temp.*) 11June15
Peel, D. K. McL. (*temp.*) 24June15

2nd Lieutenants-contd.

Banks, C. M. (*temp.*) 12July15
Dunn, H J. (*temp.*) 12July15
Davies, H. (*temp.*) 12July15
Malcolm, R. G. (*temp.*) 12July15
Thomas, M. V. (*temp.*) 12July15
Walker, H. S. (*temp.*) 12July15
Anderson, F. E. (*temp.*) 12July15
Chater, E. H. (*temp.*) 13July15
Deane, W. (*temp.*) 13July15
Baskett, E. G. (*temp.*) 13July15
Atkinson, G. FitzG. (*temp.*) 14Aug.15
Makeig-Jones, T. G. R. (*temp.*) 14Aug.15
Bishop, D. (*temp.*) 14Aug.15
Currin, R. W. (*temp.*) 14Aug.15
Davis, A. L. (*temp.*) 14Aug.15
Firth, E. H. (*temp.*) 14Aug.15
Francis, F. S. (*temp.*) 14Aug.15
Paige, R. H. (*temp.*) 14Aug.15
Parkes, M. F. M. (*temp.*) 14Aug.15
Perkins, S. M. (*temp.*) 14Aug.15
Fitt, C. W. (*temp.*) 23Aug.15
Skuce, A. (*temp.*) 26Aug.15
Percival, R. T. (*temp.*) 26Aug.15
Woodall, J. F. (*temp.*) 26Aug.15
Boyd, T. (*temp.*) 26Aug.15
Oglesby, R. H. (*temp.*) 26Aug.15

Adjutant.

Quarter-Master.

Baker, A. C., *hon. t.* (*temp.*) 23Oct.14

Attached.

2nd Lieutenants.

Osmand, G. (*temp.*) 16Sept.14
Farrer, L. H. St. G. (*temp.*) 18Nov.14
Southam, F. (*temp.*) 27Nov.14
Gray, G. G. (*temp.*) 22Dec.14
Brooks, E. W. (*temp.*) 22Dec.14
Benson, C. S. (*temp.*) 22Dec.14
Clutsom, C. A. (*temp.*) 22Dec.14
Smith, A. W. (*temp.*) 26Dec.14
Walker, T. (*temp.*) 29Dec.14
Harris, T. N. C. (*temp.*) 12Jan.15
Harris, A. D. V. (*temp.*) 12Jan.15
Prodger, S. E. (*temp.*) 14Jan.15
Copeman, H. G. H. (*temp.*) 25Jan.15
Bell, V. G. (*temp.*) 6Mar.15
Bell, R. H. (*temp.*) 26Apr.15

Appendix B

THE ROLL OF HONOUR

OF THE

OXFORDSHIRE AND BUCKINGHAMSHIRE LIGHT INFANTRY

FOR THE FIRST YEAR OF THE GREAT WAR

Being the casualties published between August 1914 *and the end of August* 1915. *All the officers are of the first and second battalions, unless otherwise specified.*

KILLED.

BAILEY, Lt. A. W., 9th Bn.
BARRINGTON-KENNETT, Sec.-Lt. A. H.
BEAUFORT, Capt. F. H.
BERLEIN, Lt. C. M., 5th Bn.
BROOKE, Capt. R. R. M.
BULL, Lt. R. E. B.
DASHWOOD, Capt. E. G., 4th Bn.
DASHWOOD, Sec.-Lt. S. A.
DAVIES, Sec.-Lt. J. T., 5th Bn.
EVELEGH, Capt. R. C.
FILLEUL, Sec.-Lt. L. A.
GIRARDOT, Sec.-Lt. P. C.
HARDEN, Capt. A. H.
HERMAN-HODGE, Sec.-Lt. J. P., 4th Bn.
HUMFREY, Lt. D. H. W., 3rd Bn., attd. 2nd. Bn.
JACKSON, Lt. J. M. H., 5th Bn.
JONES, Sec.-Lt. J.
MARSHALL, Sec.-Lt. J. S. C.
MOCKLER-FERRYMAN, Lt. H.
MURPHY, Lt. C. F.
PEPYS, Sec.-Lt. F.
RIDDLE, Sec.-Lt. F. E. L., 2nd Bn.
TURBUTT, Lt. G. M. R.
TYLDEN-PATTENSON, Sec.-Lt. A. D.
VYNER, Lt. C. J. S., 4th Bn.
WARD, Sec.-Lt. J. B. M.
WORTHINGTON, Lt. R. G.

WOUNDED

BAKER, Sec.-Lt. C. B.
BAILIE, Lt. C. W. H., attd. 2nd Rifle Brigade.
BAINES, Lt. C. S. (twice).
BARNARD, Sec.-Lt. W. L.
BLEWITT, Capt. G.

Appendix B

BIRCH, Lt. W. R., 5th Bn.
BIRD, Lt. S. H., 9th Bn.
BOARDMAN, Lt. J. H.
BOWYER, Capt. G. E. W., Bucks. Bn. (T.F.)
BRETT, Lt. R. J.
CARDY, Lt. A. G.
CARFRAE, Capt. C. F. K., 1st Bn., attd. 5th Bn.
CHIPPENDALE, Sec.-Lt. F. W. C., 3rd Bn., attd. 2nd Bn.
CLARKE, Sec.-Lt. H. F., 5th Bn.
COMBS, Lt. H. V., Bucks. Bn.
CRANMER, Lt. J. F. A., 4th Bn.
CRAWFORD, Sec.-Lt. C. B., 5th Bn.
CURRY, Sec.-Lt. P. A., 5th Bn.
DEAN, Sec.-Lt. C.
DILLON, Capt. H. M.
EDEN, Major A. J. F.
EIGHTEEN, Sec.-Lt. W. J.
FENWICK, Sec.-Lt. J. S., 5th Bn.
FOWKE, Lt. C. A. F., 3rd, attd. 2nd Bn.
GAUNTLETT, Lt. H. L., attd. 4th Bn. (T.F.)
GREEN, Lt. B., Bucks. Bn.
HAMMICK, Capt. S. F.
HENLEY, Capt. C. F.
HIGGINS, Capt. C. G.
HOBART-HAMPDEN, Sec.-Lt. G.M.H., Bucks. Bn.
HOGAN, Capt. C. J., 9th Bn., attd. K. O. Scottish Borderers (Medit. Exped. Force).
HORLEY, Sec.-Lt. C. R.
JURY, Lt. C. R., 9th Bn., attd. 5th Bn.
KIRKPATRICK, Bt.-Maj. E. H.
LARDNER, Lt. G. S., 9th Bn.
LOGAN, Capt. R. O., 5th Bn.
MEADE, Lt. J. W.
MINIFIE-HAWKINS, Sec.-Lt. S. M., 3rd Bn., attd. 2nd Bn.
NEWTON-KING, Sec.-Lt. P. F.
OWEN, Lt. R. M.
PONSONBY, Capt. A. W. N.
RADFORD, Sec.-Lt. H. D. H., 3rd Bn., attd. 1st Buffs (E. Kent Regt.).
RATTRAY, Lt. S., 9th Bn., attd. 1st Bn. K. O. Scottish Borderers (Medit. Exped. Force).
REYNOLDS, Capt. L. L. C., Bucks. Bn.
ROSE, Capt. G. K., 4th Bn.
ROSE, Capt. D. M., 4th Bn.
SCOTT, Lt. M. D.
SEWELL, Sec.-Lt. D. A. D.
SPENCER, Lt. A. V.
SWEET-ESCOTT, Sec.-Lt. F. W., 5th Bn.
TATTON, Capt. R. H. G., 3rd Bn., attd. 2nd Bn. D. of Corn. L. I.
TERRY, Lt. W. E. C.
TITHERINGTON, Lt. G. W.
TOLSON, Capt. W. C.
TUCK, Sec.-Lt. D. J., 3rd Bn., attd. 2nd Bn. D. of Corn. L. I.
WALTERS, Sec.-Lt. F. P., 3rd Bn., attd. 1st Yorkshire L.I.
WHITFELD, Lt. E. H.
WILLIAMSON, Sec.-Lt. N. B., Bucks Bn.
WINGFIELD, Lt. M. ff. R.
WOOD, Capt. H. L.
WOODING, Sec.-Lt. W.
WYLD, Lt. J. W. G.

SUFFERING FROM SHELL CONCUSSION

NEWBOLT, Lt. A. F., 3rd Bn., attd. 2nd Bn. D. of Corn. L. I.
QUINN, Lt. J. W., 3rd Bn., attd. 2nd Bn. D. of Corn. L. I.
SEWELL, Sec.-Lt. D. A. D.

Appendix B

MISSING

BUTTON, Sec.-Lt. G. T., 3rd Bn., attd. 2nd Bn. (unofficially reported wounded and prisoner).

CALDERON, Lt. G. (also wounded), 9th Bn., attd. K. O. Scottish Borderers (Medit. Exped. Force).

GODSAL, Capt. P., 3rd Bn., attd. 2nd Bn. (unofficially reported wounded and prisoner).

JOHNSTON, Lt. J. L. (also wounded 3rd Bn., attd. 2nd Bn.

RENDEL, Lt. R. D.

TUNNARD, Capt. L.

WARNER, Lt. C. J., 3rd Bn., attd. 2nd Bn.

INDEX

Index

A

CATALOGUE OF BOOKS

PUBLISHED AT THE

OFFICES OF "COUNTRY LIFE"

20, TAVISTOCK STREET, COVENT GARDEN,
LONDON, W.C.

Q

Uniform with "Small Country Houses: Their Repair and Enlargement," "Gardens for Small Country Houses," and "The House and Its Equipment."

SMALL COUNTRY HOUSES OF TO-DAY

Edited by LAWRENCE WEAVER

Large quarto, cloth, gilt, **15/-** *net;*
by post (inland) **15/6.** *Foreign and Colonial Post,* **16/6**

224 PAGES, 300 ILLUSTRATIONS

This volume fills a distinctive place, because not only is the picked work of more than forty of the best architects of the day shown by plan and photograph, but it is discussed in detail, frankly yet sympathetically. As the houses illustrated, nearly fifty in all, vary from whitewashed week-end cottages costing less than £500 to dignified country homes costing £5,000, all sorts of internal arrangement and architectural and garden treatment are brought under review. To all of moderate means who contemplate building a country house, this book will be of the utmost value.

Uniform with "Gardens for Small Country Houses," "Small Country Houses of To-day," and "Small Country Houses: Their Repair and Enlargement."

THE HOUSE AND ITS EQUIPMENT

Edited by LAWRENCE WEAVER

Large quarto, cloth, gilt, **15/-** *net;*
by post (inland) **15/6.** *Foreign and Colonial Post,* **16/6**

212 PAGES, 240 ILLUSTRATIONS

It is impossible that any one writer can deal with the many problems that arise out of the artistic and practical equipment of a house, at least with equal knowledge and sympathy. The scheme of this volume, with its forty-three chapters contributed by twenty-three experts of acknowledged ability, ensures the throwing of fresh light on scores of questions that concern the comfort and pleasure of everyone. To all who own a home, and are not wholly satisfied with it, and to all who contemplate improving an existing house or building anew, this volume will be of the utmost value.

The "Country Life" Library

OUR COMMON SEA-BIRDS

CORMORANTS, TERNS, GULLS, SKUAS, PETRELS, AND AUKS

By PERCY R. LOWE, B.A., M.B., B.C.

With Chapters by BENTLEY BEETHAM, FRANCIS HEATHERLEY, W. R. OGILVIE-GRANT, OLIVER G. PIKE, W. P. PYCRAFT, A. J. ROBERTS, etc.

Large quarto, cloth, gilt, with over 300 *pages and nearly* 250 *illustrations.* 15/- *net.* *Post free (inland)* 15/7

Unlike the majority of books dealing with birds, this volume is of interest to the general reader and to the student of ornithology alike.

It is a book that enables the reader to identify our Sea-birds by name, to understand their movements, their habits, their nests and their eggs.

The Observer says:—"We marvel at the snapshots that have been taken of birds. Every movement of their flight is now recorded; the taking off, the alighting, the swooping, the settling, the 'planing,' the struggling against the wind. And they are just the birds which the ordinary man wants to know about, because he has such opportunities of seeing them for himself on any walk along the cliff."

THE PEREGRINE FALCON AT THE EYRIE

By FRANCIS HEATHERLEY, F.R.C.S.

Illustrated with wonderful photographs by the AUTHOR and C. J. KING.

Demy quarto, cloth, gilt, 5/- *net; by inland post,* 5/6

This fascinating book on the Peregrine Falcon—the grandest bird of prey left in England—combines the salient facts of almost innumerable field notes, *written at the eyrie itself*. It is a book that should appeal with irresistible force to all true nature lovers. Many striking and unexpected facts were revealed to the author as a result of unwearying patience in a diminutive hut slung from the precipice of a lonely islet. These records are now set forth in a wonderful narrative which discloses the life history of the Peregrine Falcon from the moment of its hatching to the day it finally leaves the eyrie.

The Times says:—"We commend this faithful and truly scientific inquiry to all lovers of animals and to those who are in quest of a real knowledge of nature."

"Country Life" Library of Garden Books

THE CENTURY BOOK OF GARDENING

Edited by E. T. Cook. *A Comprehensive Work for every Lover of the Garden.* 624 *pages, with about* 600 *illustrations, many of them full-page* 4*to* (12*in. by* 8½*in.*). 21*s. net. By post* 21*s.* 10*d.*

"No department of gardening is neglected, and the illustrations of famous and beautiful gardens and of the many winsome achievements of the gardener's art are so numerous and attractive as to make the veriest cockney yearn to turn gardener. If The Century Book of Gardening does not make all who see it covet their neighbours' gardens through sheer despair of ever making for themselves such gardens as are there illustrated, it should, at any rate, inspire everyone who desires to have a garden with an ambition to make it as beautiful as he can."—*Times.*

GARDENING FOR BEGINNERS

(*A Handbook to the Garden.*) *By* E. T. Cook. *Coloured plates and over* 200 *illustrations, plans and diagrams from photographs of selected specimens of Plants, Flowers, Trees, Shrubs, Fruits, etc. Sixth Edition.* 12*s.* 6*d. net. By post,* 13*s.*

"One cannot speak in too high praise of the idea that led Mr. E. T. Cook to compile this Gardening for Beginners, and of the completeness and succinctness with which the idea has been carried out. Nothing is omitted. . . . It is a book that will be welcomed with enthusiasm in the world of gardeners."—*Morning Post.*

WALL AND WATER GARDENS

With Chapters on the Rock Garden, the Heath Garden and the Paved Water Garden. 5*th Edition. Revised and Enlarged. By* Gertrude Jekyll. *Containing instructions and hints on the cultivation of suitable plants on dry walls, rock walls, in streams, marsh pools, lakes, ponds, tanks and water margins. With* 200 *illustrations. Large* 8*vo,* 220 *pages.* 12*s.* 6*d. net. By post,* 12*s.* 11*d.*

"He who will consent to follow Miss Jekyll aright will find that under her guidance the old walls, the stone steps, the rockeries, the ponds, or streamlets of his garden will presently blossom with all kinds of flowers undreamed of, and become marvels of varied foliage."—*Times.*

COLOUR SCHEMES FOR THE FLOWER GARDEN

By Gertrude Jekyll. *With over* 100 *illustrations and planting plans. Third Edition.* 12*s.* 6*d. net. By post,* 13*s.*

"Miss Jekyll is one of the most stimulating of those who write about what may be called the pictorial side of gardening. . . . She has spent a lifetime in learning how to grow and place flowers so as to make the most beautiful and satisfying effects, and she has imparted the fruits of her experience in these delightful pages."—*Daily Mail.*

THE FRUIT GARDEN

By George Bunyard *and* Owen Thomas. 507 *pages. Size,* 10½*in. by* 7½*in.* 12*s.* 6*d. net. By post,* 13*s.*

"Without any doubt the best book of the sort yet published. There is a separate chapter for every kind of fruit, and each chapter is a book in itself—there is, in fact, everything that anyone can need or wish for in order to succeed in fruit growing. The book simply teems with illustrations, diagrams, and outlines."—*Journal of the Royal Horticultural Society.*

LILIES FOR ENGLISH GARDENS

Written and compiled by GERTRUDE JEKYLL. 8*s.* 6*d. net. By post,* 8*s.* 10*d.*

"LILIES FOR ENGLISH GARDENS is a volume in the COUNTRY LIFE Library, and it is almost sufficiently high commendation to say that the book is worthy of the journal. Miss Jekyll's aim has been to write and compile a book on Lilies which shall tell amateurs, in the plainest and simplest possible way, how most easily and successfully to grow the Lily."—*Westminster Gazette.*

THE UNHEATED GREENHOUSE

By MRS. K. L. DAVIDSON. *Cheap Edition,* 5*s. net. By post,* 5*s.* 4*d.*

"An infinity of pleasure can be obtained from the due use of an unheated house built under proper conditions, and it is the function of Mrs. Davidson's book to provide hints and directions how to build such a house, and how to cultivate the plants that can be cultivated with advantage without artificial heat."—*Pall Mall Gazette.*

THE ENGLISH VEGETABLE GARDEN

By various experts. Cheap Edition, 5*s. net. By post,* 5*s.* 6*d.*

"The book is of a thoroughly practical nature, and covers the whole ground from the trenching of the land to the gathering of the produce, and, aided by suitable illustrations, the writers have succeeded in furnishing a book which will be of inestimable advantage to the enterprising private or market gardener who would make the most of his resources."—*Field.*

CHILDREN AND GARDENS

By GERTRUDE JEKYLL. *A garden book for children, treating not only of their own little gardens and other outdoor occupations, but also of the many amusing and interesting things that occur in and about the larger home garden and near grounds. Thoroughly practical and full of pictures.* 6*s. net. By post,* 6*s.* 4*d.*

"Little bits of botany, quaint drawings of all kinds of things, pretty pictures, reminiscences and amusements—why, it is a veritable 'Swiss Family Robinson' for the bairns, and we shall be surprised and disappointed if it is not introduced into many hundreds of homes."—*Liverpool Post.*

ROCK AND WATER GARDENS: Their Making and Planting

With Chapters on Wall and Heath Gardens. By F. H. MEYER. *Edited by* E. T. COOK. 6*s. net. By post,* 6*s.* 4*d.*

"In this book the author has studied every detail of Nature's ways in order to reproduce in the garden the charms of natural scenery."—*Standard.*

TREES AND SHRUBS FOR ENGLISH GARDENS

By E. T. COOK. 12*s.* 6*d. net. By post,* 12*s.* 11*d.*

"It contains a mass of instruction and illustration not always to be found altogether when required, and as such it will be very useful as a popular handbook for amateurs and others anxious to grow trees and shrubs."—*Field.*

MY GARDEN

By EDEN PHILLPOTTS. 207 *pages.* 60 *full-page illustrations. Cheap Edition,* 6*s. net. By post,* 6*s.* 5*d.*

"It is a thoroughly practical book, addressed especially to those who, like himself, have about an acre of flower garden, and are willing and competent to help a gardener to make it as rich, as harmonious, and as enduring as possible. His chapters on irises are particularly good."—*World.*

THE SMALL ROCK GARDEN

By E. H. JENKINS. *Large Crown* 8*vo, over* 50 *illustrations and coloured frontispiece.* 2*s.* 6*d. net. By post,* 2*s.* 10*d.*

"Thoroughly practical and finely illustrated."—*Scotsman.*

A GARDEN IN VENICE

By F. EDEN. *An account of the author's beautiful garden on the Island of the Guidecca at Venice. With* 21 *collotype and* 50 *other illustrations. Parchment, limp.* 10s. 6*d. net. By post,* 10s. 10*d.*

"Written with a brightness and an infectious enthusiasm that impart interest even to technicalities, it is beautifully and rarely pictured, and its material equipment is such as to delight the lover of beautiful books."—*Glasgow Herald.*

SEASIDE PLANTING OF TREES AND SHRUBS

By ALFRED GAUT, F.R.H.S. *An interesting and instructive book dealing with a phase of arboriculture hitherto not touched upon. It is profusely illustrated, and diagrams are given explaining certain details.* 5s. *net. By post,* 5s. 4*d.*

"Mr. Gaut has accomplished a piece of very solid and extremely useful work, and one that may not be without considerable influence upon the future development of coast-side garden work and agriculture."—*Liverpool Courier.*

ROSE GROWING MADE EASY

By E. T. COOK. *A simple Rose Guide for amateurs, freely illustrated with diagrams showing ways of increasing, pruning and protecting roses.* 1s. *net. Cloth,* 1s. 6*d. net. Postage,* 3*d. extra.*

". . . Ought to be in the hands of every rose grower."—*Aberdeen Free Press.*

THE BOOK OF BRITISH FERNS

By CHAS. T. DRUERY, F.L.S., V.M.H., *President of the British Pteridological Society.* 3s. 6*d. net. By post,* 3s. 10*d.*

"The book is well and lucidly written and arranged; it is altogether beautifully got up. Mr. Druery has long been recognised as an authority on the subject."—*St. James's Gazette.*

THE HARDY FLOWER BOOK

By E. H. JENKINS. *A complete and trustworthy guide to all who are desirous of adding to their knowledge of the best means of planting and cultivating hardy flowers. Large Crown* 8vo, 50 *illustrations and coloured frontispiece.* 2s. 6*d. net. By post,* 2s. 10*d.*

"The amateur gardener who covets success should read 'The Hardy Flower Book.'"—*Daily Mail.*

THE DISEASES OF TREES

By PROFESSOR R. HARTIG. *Royal* 8*vo.* 10s. 6*d. net. By post,* 10s. 10*d.*

GARDENING MADE EASY

By E. T. COOK. *An instructive and practical gardening book of* 200 *pages and* 23 *illustrations.* 1s. *net. Cloth,* 1s. 6*d. net. Postage,* 3*d. extra.*

"The A.B.C. of Gardening."—*Scotsman.*

FRUIT GROWING FOR BEGINNERS

A simple and concise handbook on the cultivation of Fruit. By F. W. HARVEY. 1s. *net. Cloth,* 1s. 6*d. net. Postage* 3*d. extra.*

"An amazing amount of information is packed into this book."—*Evening News.*

Other Volumes in the Series of "Country Life" Military Histories

† The Royal Welsh Fusiliers
By H. AVRAY TIPPING, M.A.

† The Royal Scots — By LAWRENCE WEAVER, with a Preface by the EARL OF ROSEBERY, K.G.

† The O.T.C. and the Great War — By Captain ALAN R. HAIG-BROWN, with a Preface by Colonel Sir EDWARD WARD, Bart., K.C.B., K.C.V.O.

* The King's (Liverpool Regiment)
By T. R. THRELFALL

* The Royal Fusiliers — By OSWALD BARRON

* The King's Royal Rifles
By PROFESSOR THOMAS SECCOMBE

* The Rifle Brigade — By COLONEL WILLOUGHBY VERNER

* *6/- net; by inland post, 6/5.*
† *7/6 net; by inland post, 8/-.*

www.ingramcontent.com/pod-product-compliance
Ingram Content Group UK Ltd.
Pitfield, Milton Keynes, MK11 3LW, UK
UKHW021840270726
14058UKWH00002B/255